NONCOMMUTATIVE RINGS

By

I. N. HERSTEIN

THE
CARUS MATHEMATICAL MONOGRAPHS

Published by
THE MATHEMATICAL ASSOCIATION OF AMERICA

———

The following monographs have been published:

NONCOMMUTATIVE RINGS

By

I. N. HERSTEIN

Professor of Mathematics
University of Chicago

Published and Distributed by

THE MATHEMATICAL ASSOCIATION OF AMERICA

© 1968 by
The Mathematical Association of America

Complete Set ISBN 0-88385-000-1
Vol. 15 ISBN 0-88385-015-X

Printed in the United States of America

Current printing (last digit):

10 9 8 7 6 5

TO THE MEMORY OF MY FATHER

PREFACE

This book is not intended as a treatise on ring theory. Instead, the intent here is to present a certain cross-section of ideas, techniques and results that will give the reader some inkling of what is going on and what has gone on in that part of algebra which concerns itself with noncommutative rings. There are many portions of great importance in the theory which are not touched upon or which are merely mentioned in passing. On the other hand there is a rather detailed treatment given to some aspects of the subject.

While the account given here is not completely self-contained, to follow it does not require a great deal beyond a good first course in algebra. Perhaps I should spell out what I would expect in such a course. To begin with one should have been introduced to some of the basic structures of algebra—groups, rings, fields, vector spaces—and to have seen some of the basic theorems about them. One would want a good familiarity with homomorphisms, the early homomorphism theorems, quotient structures and the like. One should have learned with some thoroughness linear algebra—the fundamental theorems about linear transformations on a vector space. This type of material can be found in many books, for instance, Birkhoff and MacLane *A Survey of Modern Algebra* or my book *Topics in Algebra*.

Beyond these standard topics cited above I shall make frequent use of results from the theory of fields. All these can be found in the chapter on field theory in van der Waerden's *Modern Algebra*. My advice, to the reader not familiar with this material, is to read into a proof until such a result is cited and then to read about the notions arising in van der Waerden's book. Finally, I shall continually use Zorn's Lemma and the axiom of choice.

A great deal of what is done in this book is based on selected parts of two sets of my notes published in the University of Chicago lecture notes series. Part of this selection and weeding process, polishing and blending together was accomplished in a course I gave at Bowdoin College, under the auspices of the Mathematical Association of America, in the summer of 1965 to a group of mathematicians teaching at various colleges and smaller universities. I should like to thank the participants in that course for their patience and enthusiasm. There are many others I should like to thank, Nathan Jacobson and Irving Kaplansky, for the part they and their work have played in my formation as a mathematician, Shimshon Amitsur for the many pleasant hours spent together working and discussing ring theory and my students, Claudio Procesi and Lance Small, for taking the notes at Bowdoin and for their stimulating comments, suggestions and improvements.

CONTENTS

CHAPTER 1

THE JACOBSON RADICAL

This chapter has as its major goal the creation of the first steps needed to construct a general structure theory for associative rings. The aim of any structure theory is the description of some general objects in terms of some simpler ones—simpler in some perceptible sense, perhaps in terms of concreteness, perhaps in terms of tractability. Of essential importance, after one has decided upon these simpler objects, is to find a method of passing down to them and to discover how they weave together to yield the general system with which we began.

In carrying out such a program there are many paths one can follow, many classes of candidates for these simpler objects, and one must choose among these for that theory which is most fruitful in producing decisive results. In the case of rings there seems to be no doubt that the fundamental structure theory laid out by Jacobson is the appropriate one. The best proof of this remark is the host of striking theorems which have resulted from the use of these methods.

1. Modules. Essential to everything that we shall discuss—in fact essential in every phase of algebra—is the notion of a module over a ring R or, in short, an R-module. To be absolutely precise we should say a right R-module for we shall allow the elements of R to act on the module from the right. However we shall merely say R-module, understanding by that term a right R-module. Briefly an R-module is a vector space over a ring R; more formally,

DEFINITION. *The additive abelian group M is said to be an R-module if there is a mapping from $M \times R$ to M (sending (m, r) to mr) such that:*

1. $m(a+b) = ma + mb$
2. $(m_1 + m_2)a = m_1a + m_2a$
3. $(ma)b = m(ab)$

for all $m \in M$ and all $a, b \in R$.

If R should have a unit element, 1, and if $m1 = m$ for all $m \in M$ we then describe M to be a *unitary* R-module.

Note that the definition made above merely says that the ring elements induce endomorphisms on M considered merely as an additive abelian group and that furthermore these endomorphisms induced behave as they should with respect to the addition and multiplication of such endomorphisms. More succinctly put, R is homomorphically imbedded in the ring of all endomorphisms of the additive group of M. For unitary modules we impose the further condition that this imbedding respect the unit element of R, that is, that it correspond to the identity endomorphism.

Mathematics abounds with examples of modules; we shall limit ourselves to two examples for the moment, constructed intrinsically from R itself.

Let R be any ring and let ρ be a right ideal of R. We impose on ρ a natural R-module structure by defining the action of R on ρ to coincide with the product of elements in R. That ρ is an R-module is nothing but a restatement of the fact that it is a right ideal of R.

Using ρ we can construct yet another R-module; let R/ρ be the quotient group of R by ρ considered as additive groups, that is, R/ρ consists of the cosets $x + \rho$ where x ranges over R. Of course R/ρ is not in general a ring—for this to be true ρ would have to be something more, namely a two-sided ideal of R—but it does at least carry the structure of an R-module. We achieve this by defining $(x+\rho)r \equiv xr+\rho$ for all $x+\rho \in R/\rho$ and all $r \in R$. Since ρ is a right ideal of R this definition of the module action makes sense; the verification of the various module axioms is a routine triviality.

Of course vector spaces over fields are examples of modules, in fact of very nice modules. There we have that only the zero element of the field can annihilate a nonzero vector. For a module M over an arbitrary ring R this may be far from true, indeed it is quite possible to have $Mr=(0)$ for some $r \neq 0$ in R. The situation in which this cannot happen is, in some sense, a decent one and we single it out. We say that M is a *faithful R-module* (or that R acts *faithfully* on M) if $Mr=(0)$ forces $r=0$. We now set up a measure of the lack of fidelity of R on M.

Definition. *If M is an R-module then $A(M) = \{x \in R \mid Mx=(0)\}$.*

Lemma 1.1.1. *$A(M)$ is a two-sided ideal of R. Moreover, M is a faithful $R/A(M)$-module.*

Proof. That $A(M)$ is a right ideal of R is immediate from the axioms for an R-module. To see that it is also a left ideal we proceed as follows: if $r \in R$ and $a \in A(M)$ then $M(ra) = (Mr)a \subset Ma \subset (0)$, hence $ra \in A(M)$. This proves that $A(M)$ is a two-sided ideal of R.

We now make of M an $R/A(M)$-module by defining, for $m \in M$, $r+A(M) \in R/A(M)$, the action $m(r+A(M)) = mr$. If $r+A(M) = r'+A(M)$ then $r-r' \in A(M)$ hence $m(r-r') = 0$ for all $m \in M$, that is to say, $mr \equiv mr'$. This in its turn tells us that $m(r+A(M)) = mr = mr' = m(r'+A(M))$; the action of $R/A(M)$ on M has been shown to be well defined. The verification that this defines the structure of an $R/A(M)$-module on M we leave to the reader. Finally, to see that M is a faithful $R/A(M)$-module we note that if $m(r+A(M))=0$ for all $m \in M$ then by definition $mr=0$ hence $r \in A(M)$. This says that only the zero element of $R/A(M)$ annihilates all of M.

We formalize some remarks made earlier. Let M be an R-module; for $a \in R$ we define T_a: $M \to M$ by $mT_a = ma$ for all $m \in M$. Since M is an R-module T_a is an endomorphism of the additive group of M, that is, $(m_1+m_2)T_a = m_1T_a + m_2T_a$ for all m_1, $m_2 \in M$. Let $E(M)$ be the set of all endomorphisms of the additive group of M; defining, as usual, for ϕ, $\psi \in E(M)$ the sum $\phi + \psi$ by $m(\phi+\psi) = m\phi + m\psi$ and the product $\phi\psi$ by $m(\phi\psi) = (m\phi)\psi$ we see that $E(M)$ is a ring.

Consider the mapping Φ: $R \to E(M)$ define by $\Phi(a) = T_a$. Going back to the definition of an R-module we see that $\Phi(a+b) = \Phi(a) + \Phi(b)$ and $\Phi(ab) = \Phi(a)\Phi(b)$, in short Φ is a ring homomorphism of R into $E(M)$. What is Ker Φ, the kernel of Φ? Clearly if $a \in A(M)$ then $Ma = (0)$ hence $0 = T_a = \Phi(a)$, that is, $a \in$ Ker Φ. On the other hand if $a \in$ Ker Φ then $T_a = 0$ leading to $Ma = MT_a = (0)$, that is, $a \in A(M)$. Therefore the image of R in $E(M)$ is isomorphic to $R/A(M)$. We have proved

LEMMA 1.1.2. $R/A(M)$ *is isomorphic to a subring of* $E(M)$.

In particular if M is a faithful R-module, one for which $A(M) = (0)$, this lemma says that we may consider R as a subring of the ring of endomorphisms of M as an additive group, and so as some ring of endomorphisms of M.

From the interrelation of the R-module M with R we have produced certain elements, the T_a as a ranges over R, in $E(M)$. How do these elements sit in $E(M)$? To be more precise, what elements in $E(M)$ commute with all these T_a?

DEFINITION. *The commuting ring of R on M is $C(M)$ $= \{ \psi \in E(M) \mid T_a \psi = \psi T_a \text{ all } a \in R \}$.*

$C(M)$ is certainly a subring of $E(M)$. If $\psi \in C(M)$ then for any $m \in M$ and $a \in R$

$$(m\psi)a = (m\psi)T_a = m(\psi T_a) = m(T_a\psi) = (mT_a)\psi = (ma)\psi,$$

that is, ψ is not only an endomorphism of M as an additive group but is in fact a homomorphism of M into itself as an R-module. We have identified $C(M)$ as the ring of all *module* endomorphisms of M.

Without going into the matter in detail it is clear what one means by a submodule, quotient module, homomorphism of modules. It is equally clear that the usual homomorphism theorems carry over in their entirety from vector spaces to our present context. We single out a special kind of R-module.

DEFINITION. *M is said to be an irreducible R-module if $MR \neq (0)$ and if the only submodules of M are (0) and M.*

For an irreducible R-module M the commuting ring turns out to be rather special. This is the content of an old and basic result known as *Schur's Lemma*.

THEOREM 1.1.1. *If M is an irreducible R-module then $C(M)$ is a division ring.*

Proof. To prove the theorem all we must do is show that any nonzero element in $C(M)$ has an inverse in $C(M)$. Actually we really need but show that if $\theta \neq 0 \in C(M)$ then θ is invertible in $E(M)$. For if $\theta^{-1} \in E(M)$ then from $\theta T_a = T_a \theta$ we immediately have that $T_a \theta^{-1} = \theta^{-1} T_a$, forcing θ^{-1} to be in $C(M)$.

Suppose that $\theta \neq 0 \in C(M)$; if $W = M\theta$ then for all $r \in R$, $Wr = WT_r = (M\theta)T_r = (MT_r)\theta \subset M\theta = W$. Consequently W is a submodule of M. Since $\theta \neq 0$ by the irreducibility of M we deduce that $W\theta = M$ or, in other words, that θ is an onto mapping.

We claim that θ is a monomorphism for, as is quickly verified, Ker θ is a submodule of M and is not all of M since $\theta \neq 0$. Thus Ker $\theta = (0)$. θ being both surjective and a monomorphism we obtain that θ^{-1} exists in $E(M)$. With this Schur's Lemma has been proved.

We pause to look at Schur's Lemma in some very particular contexts.

Let F be a field and let F_n be the ring of all $n \times n$ matrices over F. We consider F_n as the ring of all linear transformations on the vector space V of n-tuples of elements of F. If A is a subset of F_n let $\overline{A}$ be the subalgebra generated by A over F. Clearly V is a faithful F_n, and so, $\overline{A}$-module. V is, in addition, both a unitary and irreducible F_n-module.

We say the set of matrices A is irreducible if V is an irreducible $\overline{A}$-module. In matrix terms this merely says that there is no invertible matrix S in F_n so that

$$S^{-1}aS = \left(\begin{array}{c|c} a_1 & 0 \\ \hline * & a_2 \end{array} \right)$$

for all $a \in A$. The commuting ring of A (that is, of $\overline{A}$) on V is merely the set of all matrices in F_n that commute with all elements of A.

In case F is an algebraically closed field the only division ring having F in its centers and finite dimensional over F, is F itself (we shall see this later). Hence in this particular case the only matrices which commute with an irreducible set of matrices must be the scalars. This is the classical form of Schur's Lemma.

We specialize this discussion to two interesting cases.

1. Let F be the field of real numbers. In F_2 consider the matrix

$$a = \begin{pmatrix} 0 & -1 \\ 1 & 0 \end{pmatrix};$$

since a has no real characteristic roots we see that $A = \{a\}$ is irreducible. What is the commuting ring of A? From

$$\begin{pmatrix} \alpha & \beta \\ \gamma & \delta \end{pmatrix}\begin{pmatrix} 0 & -1 \\ 1 & 0 \end{pmatrix} = \begin{pmatrix} 0 & -1 \\ 1 & 0 \end{pmatrix}\begin{pmatrix} \alpha & \beta \\ \gamma & \delta \end{pmatrix}$$

we obtain $\alpha = \delta$, $\beta = -\gamma$ hence the commuting ring of A is the set of all matrices

$$\left\{ \begin{pmatrix} \alpha & -\beta \\ \beta & \alpha \end{pmatrix} \right\},$$

a field isomorphic to the complex numbers.

2. Again let F be the field of real numbers. In F_4 we consider the two matrices

$$a = \begin{pmatrix} 0 & -1 & 0 & 0 \\ 1 & 0 & 0 & 0 \\ 0 & 0 & 0 & -1 \\ 0 & 0 & 1 & 0 \end{pmatrix}; \qquad b = \begin{pmatrix} 0 & 0 & -1 & 0 \\ 0 & 0 & 0 & 1 \\ 1 & 0 & 0 & 0 \\ 0 & -1 & 0 & 0 \end{pmatrix}$$

We leave it to the reader to verify that $A = \{a, b\}$ is irreducible and that the commuting ring of A is the set of all 4×4 real matrices of the form

$$\left\{ \begin{pmatrix} \alpha & -\beta & -\gamma & -\delta \\ \beta & \alpha & \delta & -\gamma \\ \gamma & -\delta & \alpha & \beta \\ \delta & \gamma & -\beta & \alpha \end{pmatrix} \right\}.$$

This is a 4-dimensional division ring over the real field which is isomorphic to the real quaternions.

We close this section with an intrinsic description of all the irreducible modules of a given ring R.

LEMMA 1.1.3. *If M is an irreducible R-module then M is isomorphic as a module to R/ρ for some maximal right ideal ρ of R. Moreover there is an $a \in R$ such that $x - ax \in \rho$ for all $x \in R$. Conversely, for every such maximal right ideal ρ of R, R/ρ is an irreducible R-module.*

Proof. Since M is irreducible, by the very definition we must have that $MR \neq (0)$. Since $S = \{ u \in M \mid uR = (0) \}$ is a submodule of M and is not M it must be the (0). Equivalently, if $m \neq 0$ is in M then $mR \neq (0)$. However, mR is a submodule of M hence it must be all of M. Define $\Psi : R \to M$ by $\Psi(r) = mr$ for every $r \in R$. We see at once that Ψ is a homomorphism of R into M as R-modules; since $mR = M$ we have that Ψ is surjective. Finally, $\operatorname{Ker} \Psi = \{ x \in R \mid mx = 0 \}$ is a right ideal ρ; by a standard homomorphism theorem we have that M is isomorphic to R/ρ as an R-module.

Any right ideal of R which properly contains ρ maps, under ϕ, into a submodule of M. Hence ρ is a maximal right ideal of R. We now produce the desired element a in R. Since $mR = M$ there is an element $a \in R$ such that $ma = m$. Therefore for any $x \in R$ $max = mx$, which is to say $m(x - ax) = 0$. This puts $x - ax$ in ρ.

We leave the proof of the converse to the reader.

2. The radical of a ring. In setting up a structure

theory for a category of algebraic objects it is desirable to be able to recognize what special classes of objects are "nice" and to be able to measure the lack of "niceness" in the general object of this category. One then wants some sort of passage from the general object to these better-behaved ones.

It is towards this goal that we introduce the radical of a ring. We shall see that a ring having (0) as radical has a rather concrete description in terms of particular and often more manageable rings. Furthermore, any ring modulo its radical will turn out to be a ring having (0) as radical. In order to study a general ring, then, we want to slice out of the ring a certain piece—the so-called radical—in such a way that we do not slice out too much, so that the piece being cut away is capable of description yet at the same time we do not want to cut out too little, so that the object resulting after the excision is also capable of description.

The motivation for the definition we make comes primarily from the representation theory of groups. In the classical theory of finite dimensional algebra the radical was defined in a completely different way. As we shall later see our definition will coincide with the classical one in the classical situations.

DEFINITION. *The radical of R, written as $J(R)$, is the set of all elements of R which annihilate all the irreducible R-modules. If R has no irreducible modules we put $J(R) = R$.*

There are many possible radicals; the one we defined and shall use throughout is often called the *Jacobson radical*.

Note that $J(R) = \cap A(M)$ where this intersection runs over all irreducible R-modules M. Since the $A(M)$ are two-sided ideals of R we see that $J(R)$ is a two-sided ideal of R. In all fairness we should call $J(R)$ the right radical of R for it has been defined in terms of right R-modules. We could similarly define a left radical. Fortunately these two turn out to be the same so we shall be spared making such a left-right distinction.

In Lemma 1.1.3 we saw that every irreducible R-module arises as R/ρ where ρ is a maximal right ideal enjoying one further property, namely, the existence of some element $a \in R$ such that $x-ax \in \rho$ for all $x \in R$. This motivates the

DEFINITION. *A right ideal ρ of R is said to be regular if there is an $a \in R$ such that $x - ax \in \rho$ for all $x \in R$.*

If R has a unit element (in fact, a left unit will do) then all its right ideals are regular.

DEFINITION. *If ρ is a right ideal of R then $(\rho : R) = \{x \in R \mid Rx \subset \rho\}$.*

Let ρ be a maximal right ideal of R which is also assumed to be regular and let $M = R/\rho$. What is $A(M)$? If $x \in A(M)$ then $Mx = (0)$, which is to say, $(r+\rho)x = \rho$ for all $r \in R$. This latter says that $Rx \subset \rho$, hence $A(M) \subset (\rho : R)$. Similarly $(\rho : R) \subset A(M)$ whence $A(M) = (\rho : R)$. Since ρ is regular there is an $a \in R$ with $x - ax \in \rho$ for all $x \in R$; in particular, if $x \in (\rho : R)$ then since $ax \in Rx \subset \rho$ we get $x \in \rho$. We thus see that $A(M) = (\rho : R)$ is the largest two-sided ideal of R which lies in ρ. In view of Lemma 1.1.3 we can say

THEOREM 1.2.1. *$J(R) = \cap(\rho : R)$ where ρ runs over all the regular maximal right ideals of R, and where $(\rho : R)$ is the largest two-sided ideal of R lying in ρ.*

We want to sharpen this. To do so we need a preliminary result.

LEMMA 1.2.1. *If ρ is a regular right ideal of R then ρ can be imbedded in a maximal right ideal of R which is regular.*

Proof. Let $a \in R$ be such that $x - ax \in \rho$ for all $x \in R$. Clearly $a \notin \rho$ otherwise $ax \in \rho$ hence $x \in \rho$ for all $x \in R$ leading to $\rho = R$.

Let $\mathfrak{M}$ be the set of all proper right ideals of R which contain ρ. If $\rho' \in \mathfrak{M}$ then $a \notin \rho'$ for otherwise $x - ax \in \rho \subset \rho'$ would yield $\rho' = R$. We can therefore apply Zorn's lemma to $\mathfrak{M}$ to obtain a ρ_0 which is maximal in $\mathfrak{M}$. ρ_0 is certainly regular for $x - ax \in \rho \subset \rho_0$. Moreover, ρ_0 is a maximal right ideal of R. This establishes the lemma.

The lemma removes all linguistic ambiguities from the phrase, maximal regular right ideal, a phrase we shall often use.

We are now in a position to characterize the radical of R in terms of the maximal right ideals themselves rather than by means of the $(\rho : R)$ of Theorem 1.2.1. This is

Theorem 1.2.2. $J(R) = \cap \rho$ *where ρ runs over all the maximal regular right ideals of R.*

Proof. By Theorem 1.2.1 $J(R) = \cap (\rho : R)$ and since $(\rho : R) \subset \rho$ we have that $J(R) \subset \cap \rho$ where ρ varies over the maximal regular right ideals of R.

For the other direction let $\tau = \cap \rho$ and let $x \in \tau$. We claim that $\{xy + y \,|\, y \in R\}$ is all of R. If not, being a regular right ideal (with $a = -x$), by Lemma 1.2.1 it would be contained in a maximal regular right ideal ρ_0. Since $x \in \cap \rho$ we have $xy \in \rho_0$ and so $y \in \rho_0$ for all $y \in R$, a contradiction. Hence $\{xy + y \,|\, y \in R\} = R$; in particular, for some $w \in R$, $-x = w + xw$, that is, $x + w + xw = 0$. If $\tau \not\subset J(R)$ then for some irreducible R-module M we must have $M\tau \neq (0)$, and so $m\tau \neq (0)$ for some $m \in M$. As a nonzero submodule of M we get that $m\tau = M$. Thus for some $t \in \tau$, $mt = -m$; since $t \in \tau$ we have seen above that there is an $s \in R$ such that $t + s + ts = 0$. Computing

$0 = m(s+t+ts) = ms+mt+mts = ms-m-ms = -m$ we get the contradiction $m = 0$. Hence $M\tau = (0)$ for all irreducible R-modules M thereby placing in $J(R)$. This completes the proof of the theorem.

In the course of the second half of the last proof we showed that every $x \in \cap \rho = J(R)$ has a companion element $y \in R$ such that $x+y+xy = 0$. We seek to characterize $J(R)$ by means of this formal property. As a matter of fact, in Jacobson's original paper, where this material was first thoroughly developed, he used this approach to the subject matter.

DEFINITION. *An element $a \in R$ is said to be right-quasi-regular if there is an $a' \in R$ such that $a+a'+aa' = 0$. We call a' a right-quasi-inverse of a.*

We can similarly define *left-quasi-regularity.* Note that if R should have a unit element 1 then a is right-quasi-regular if and only if $1+a$ is right invertible in R. We say that a *right ideal* of R is *right-quasi-regular* if each of its elements is.

In the course of proving Theorem 1.2.2 we actually showed two additional facts:

1. $J(R)$ is right-quasi-regular.

2. If ρ is a right-quasi-regular right ideal of R then $\rho \subset J(R)$ (the proof that $\tau \subset J(R)$ given in Theorem 1.2.2 carries over verbatim to ρ). Hence we already have

THEOREM 1.2.3. $J(R)$ *is a right-quasi-regular right ideal of R and contains all the right-quasi-regular right ideals of R; that is, $J(R)$ is the unique maximal right-quasi-regular right ideal of R.*

Suppose that an element a in R is both left and right-quasi-regular. Hence there are b, $c \in R$ such that $a+b+ba = 0$ and $a+c+ac = 0$. Therefore $ba+bc+bac$

$=0$ and $ac+bc+bac=0$; this yields that $ba=ac$. In conjunction with $a+b+ba=a+c+ac=0$ we get $b=c$. In other words the left and right quasi-inverses of a are the same.

Suppose now that $a \in J(R)$; thus there is an $a' \in R$ such that $a+a'+aa'=0$. Because $a'=-a-aa'$ and $a \in J(R)$ we get that $a' \in J(R)$. In view of this there is an element $a'' \in J(R)$ such that $a'+a''+a'a''=0$. Thus a' has a as left-quasi-inverse and a'' as right-quasi-inverse; by the remark made above $a=a''$. This merely says that $a+a'+a'a=0$ or, equivalently, that a is left-quasi-regular. We have shown that $J(R)$ is a left-quasi-regular ideal of R.

This allows us to dispose of the annoying possibility of the existence of distinct right and left radicals. Using the left analog of Theorem 1.2.3 the right radical of R, as a left-quasi-regular ideal of R, is contained in the left radical. Similarly, the left radical is contained in the right one. In short, these two notions of radical coincide and we no longer need worry about it. We point out that as a consequence of this the intersection of the maximal regular right ideals of a ring equals that of the maximal regular left ideals.

In some circumstances a right ideal can be shown to be right-quasi-regular by explicitly exhibiting the quasi-inverses for its elements.

DEFINITION. (a) *We say that an element $a \in R$ is nilpotent if $a^m=0$ for some integer n.* (b) *We say that a right (left, two-sided) ideal is nil if each of its elements is nilpotent.* (c) *We say that a right (left, two-sided) ideal ρ is nilpotent if there is an integer m such that $a_1 a_2 \cdots a_m=0$ for all $a_1, \cdots, a_m \in \rho$.*

If I, J are two right (left, two-sided) ideals of R we denote by IJ the additive subgroup of R generated by all products ab where $a \in I$, $b \in J$. Then IJ is a right (left, two-sided) ideal of R. By induction we define $I^n = I^{n-1}I$ where $I^1 = I$. The right ideal ρ is *nilpotent*, in these terms, if $\rho^m = (0)$ for some m.

While a nilpotent right ideal is nil a nil one, in general, need not be nilpotent. There are many interesting theorems in ring theory, whose conclusion is precisely this, that under certain hypotheses on the ring a nil ideal is rendered nilpotent. We shall see some theorems of this nature in this book.

Suppose that $a^m = 0$; if

$$b = -a + a^2 - a^3 + \cdots + (-1)^{m-1}a^{m-1},$$

a simple computation reveals that $a + b + ab = 0$. In consequence of this every nil right ideal is right-quasi-regular. In view of Theorem 1.2.3 we have

LEMMA 1.2.2. *Every nil right or left ideal of R is contained in $J(R)$.*

Let us recall that an *algebra A* over a field F is a vector space over F in which one has a product such that relative to this product A is a ring and such that the vector space structure is interlocked with the ring structure by means of the rule $\alpha(ab) = (\alpha a)b = a(\alpha b)$ for all a, $b \in A$ and $\alpha \in F$. If A should have a unit element 1 this says that the scalars $F1$ lie in the center of A. With or without a unit element it says that the mappings $T_a: A \to A$ defined by $xT_a = xa$ and $L_a: A \to A$ defined by $xL_a = ax$ are linear transformations with respect to F.

For an algebra A it is natural to define ideals, homomorphisms, etc. by insisting that all the structures on A —that is, as a vector space and as a ring—are respected. Using this we could carry out everything we did to get a radical of A *as an algebra*. It is conceivable that this differs from the radical of A as a ring. We now show that in point of fact they are the same.

Let ρ be a maximal regular right ideal of A considered as a ring. We claim that ρ is automatically a subspace of A over F. If not, $F\rho \not\subset F$; however, from the defining properties of an algebra $F\rho$ is a right ideal of R. By the maximality of ρ we conclude that $A = F\rho + \rho$. Thus $A^2 = (F\rho + \rho)A \subset (F\rho)A + \rho A \subset \rho(FA) + \rho A \subset \rho$. Since ρ is regular there is an $a \in A$ such that $x - ax \in \rho$ for all $x \in A$; but $ax \in A^2 \subset \rho$ leaving us with the contradiction that $\rho = A$. Having now established that every maximal regular right ideal of A as a ring is a maximal regular right ideal of A as an algebra we conclude, on invoking Theorem 1.2.2, that the algebra radical of A coincides with the ring radical of A.

A natural question immediately presents itself: if we factor out the radical of R what is the radical of the resulting ring? The answer is simple and concise as we see in

THEOREM 1.2.4. $J(R/J(R)) = (0)$.

Proof. Let $\overline{R} = R/J(R)$ and let ρ be a maximal regular right ideal of R. We know that $\rho \supset J(R)$ therefore by the usual homomorphism theorems $\bar{\rho} = \rho/J(R)$ is a maximal right ideal of $\overline{R}$. It is regular for if $x - ax \in \rho$ then $\bar{x} - \bar{a}\bar{x} \in \bar{\rho}$ for all $\bar{x} \in \overline{R}$. Since $J(R) = \cap \rho$, ρ running over the maximal regular right ideals of R we get $(0) = \cap \bar{\rho}$. Hence from Theorem 1.2.2 $J(\overline{R}) \subset \cap$ (all maximal regular right ideals of $\overline{R}$) $\subset \cap \bar{\rho} = (0)$, which is the desired result.

The property of the radical exhibited in Theorem 1.2.4 is one of a set of "radical-like" properties. Studies of such general radical properties have been made by Amitsur and Kurosh. Two other such results explore the nature of the radicals of ideals of R and of matrix rings over R. We dispose of these in the next two theorems but, first, a definition.

DEFINITION. *R is said to be semi-simple if* $J(R) = (0)$.

In these terms Theorem 1.2.4 can be read as $R/J(R)$ is semisimple for any ring R. In order to avoid the heaviness of the phrase two-sided ideal (which would appear often) we shall henceforth use the word "ideal" unadorned with adjectives to mean a two-sided ideal.

THEOREM 1.2.5. *If A is an ideal of R then $J(A)$* $A \cap J(R)$.

Proof. If $a \in A \cap J(R)$ then as an element of $J(R)$, a is right-quasi-regular. Its quasi-inverse $a' = a - aa'$ is thus in A since A is an ideal of R. In short, $A \cap J(R)$ is a quasi-regular ideal of A so must be contained in $J(A)$ by Theorem 1.2.3.

Suppose now that ρ is a maximal regular right ideal of R and let $\rho_A = A \cap \rho$. If $A \not\subset \rho$ the maximality of ρ forces $= A + \rho = R$ therefore

$$ R/\rho \approx \frac{A + \rho}{\rho} \approx \frac{A}{A \cap \rho} = A/\rho_A. $$

Since R/ρ is irreducible we get that ρ_A is a maximal right ideal of A. Since ρ is regular $x - bx \in \rho$ for some $b \in R$; $b = a + r$ with $a \in A$, $r \in \rho$. Hence $\rho \ni x - bx = x - (a + r)x = x - ax - rx$ giving us that $x - ax \in \rho$. In particular we see that ρ_A is regular in A. Therefore $J(A) \subset \rho_A$ for all maximal regular right ideals ρ of R which do not contain A and certainly also for those which do. In other words, $J(A) \subset \cap \rho_A = (\cap \rho) \cap A = J(R) \cap A$.

The two opposite containing relations give us the desired result $J(A) = A \cap J(R)$.

COROLLARY. *If R is semisimple then so is every ideal of R.*

The result is false if we merely assume A to be a one-sided ideal. For instance, let R be the ring of all 2×2 matrices over a field F. It is easy to show that R has no nontrivial ideals and that $J(R) = (0)$. Let

$$\rho = \left\{ \begin{pmatrix} \alpha \beta \\ 0 0 \end{pmatrix} \mid \alpha, \beta \in F \right\} ;$$

ρ is a right ideal of R and a quick check reveals that

$$\begin{pmatrix} 0 \beta \\ 0 0 \end{pmatrix} \in J(\rho),$$

hence $J(\rho) \neq (0) = \rho \cap J(R)$.

Another basic "radical-like" property is the behavior of the radical when one passes from a ring to the ring of matrices over this ring. If S is a ring let S_m denote the ring of all $m \times m$ matrices over S. The precise result is

THEOREM 1.2.6. $J(R_m) = J(R)_m$.

Proof. Let M be an irreducible R-module;

$$M^{(m)} = \left\{ (m_1, \cdots, m_m) \mid m_i \in M \right\}$$

is an R_m-module in the natural way. Moreover, as we leave to the reader to verify, it is an irreducible R_m-module. If $(\alpha_{ij}) \in J(R_m)$ then for all $m_i \in M$, $(m_1, \cdots, m_m)(\alpha_{ij}) = (0, \cdots, 0)$; this yields that $M\alpha_{ij} = (0)$ for all i, j and so $\alpha_{ij} \in J(R)$. We have shown that $J(R_n) \subset J(R)_n$.

We now wish to show that $J(R)_n \subset J(R_n)$; we do it by showing that it is a quasi-regular ideal of R_n. Consider

$$\rho_1 = \left\{ \begin{pmatrix} \alpha_{11} & \alpha_{12} & \cdots & \alpha_{1n} \\ 0 & 0 & \cdots & 0 \\ \cdot & \cdot & \cdots & \cdot \\ 0 & 0 & \cdots & 0 \end{pmatrix} \middle| \alpha_{1j} \in J(R) \right\} ;$$

ρ_1 is a right ideal of R_n. If

$$X = \begin{pmatrix} \alpha_{11} & \alpha_{12} & \cdots & \alpha_{1n} \\ 0 & 0 & \cdots & 0 \\ . & . & \cdots & . \\ 0 & 0 & \cdots & 0 \end{pmatrix} \in \rho_1 \quad \text{let } Y = \begin{pmatrix} \alpha_{11}' & 0 & \cdots & 0 \\ 0 & 0 & \cdots & 0 \\ . & & \cdots & . \\ 0 & 0 & \cdots & 0 \end{pmatrix}$$

where $\alpha_{11} + \alpha_{11}' + \alpha_{11}\alpha_{11}' = 0$. Computing $W = X + Y + XY$ we get a strictly triangular matrix hence $W^n = 0$. Therefore W is right-quasi-regular. If $W + Z + WZ = 0$ on carrying out the required calculation we see that $X + (Y + Z + YZ) + X(Y + Z + YZ) = 0$. ρ_1 has been shown to be a right quasi-regular right ideal of R_n so is in $J(R_n)$. We similarly prove that

$$\rho_i = \left\{ \begin{pmatrix} 0 & \cdots & 0 \\ 0 & \cdots & 0 \\ \alpha_{i1} & \cdots & \alpha_{in} \\ 0 & \cdots & 0 \end{pmatrix} \middle| \alpha_{ij} \in J(R) \right\} \quad \text{is in } J(R_n).$$

Since $J(R_n)$ is closed under addition we get that $\rho_1 + \cdots + \rho_n \subset J(R_n)$ hence $J(R)_n \subset J(R_n)$. This completes the proof.

3. Artinian rings. We now shall consider the implications and meaning of some of the material already developed in a classical framework studied earlier in ring theory.

DEFINITION. *A ring is said to be right Artinian if any non-empty set of right ideals has a minimal element.*

For the sake of brevity we shall drop the "right" in

right Artinian and refer to them as merely Artinian rings. Artinian rings may be equivalently defined in terms of descending chains of right ideals. This is the content of the

REMARK. A ring is Artinian if and only if any descending chain of right ideals of R, $\rho_1 \supset \rho_2 \supset \cdots \supset \rho_m \supset \cdots$ becomes stationary, that is, if from some point on all the ρ_i's are equal.

To see this let $\rho_1 \supset \cdots \supset \rho_m \supset \cdots$ be a descending chain of right ideals. If R is Artinian the set $\mathfrak{M} = \{\rho_i\}$ of right ideals has a minimal element ρ_m. But then, since for $i > m$ $\rho_m \supset \rho_i$ and since ρ_m is minimal in $\mathfrak{M}$, we get $\rho_i = \rho_m$.

On the other hand suppose that R satisfies the given chain condition on right ideals. Let $\mathfrak{M}$ be a nonempty set of right ideals of R. Pick $\rho_1 \in \mathfrak{M}$; either it is minimal in $\mathfrak{M}$ or it is not. If not, $\rho_1 \supset \rho_2$ for some $\rho_2 \in \mathfrak{M}$. Again, either ρ_2 is minimal in $\mathfrak{M}$ or it is not. This way we get a chain of $\rho_1 \supset \rho_2 \supset \cdots \supset \rho_m \supset \cdots$. Since by assumption this chain becomes stationary we get a minimal element in $\mathfrak{M}$. Whence R is Artinian.

Note that we have used the axiom of choice in demonstrating this equivalent formulation of the notion of an Artinian ring.

We pause to look at some examples of Artinian rings:

1. Any division ring is Artinian—in fact it has no nontrivial right ideals.

2. The ring of $n \times n$ matrices over a division ring is Artinian. We leave to the reader the following—it is not trivial—if R is Artinian then so is R_n.

3. A finite direct sum of Artinian rings is again Artinian.

-4. If A is a finite dimensional algebra over a field then A is Artinian *as an algebra*, that is, where we mean by a right ideal an algebra right ideal. Although Artinian as an algebra it need not be Artinian as a ring. A simple example of this is the ring $A = Fu$, F the field of rational numbers and $u^2 = 0$.

5. A homomorphic image of an Artinian ring is Artinian.

For Artinian rings the radical is very special. This is

THEOREM 1.3.1. *If R is Artinian then $J(R)$ is a nilpotent ideal.*

Proof. Let $J = J(R)$; consider the descending chain of right ideals $J \supset J^2 \supset \cdots \supset J^n \supset \cdots$. Since R is Artinian there is an integer n such that $J^n = J^{n+1} = \cdots = J^{2n} = \cdots$. Hence if $xJ^{2n} = (0)$ then $xJ^n = 0$.

We want to prove that $J^n = (0)$; suppose it is not. Let $W = \{x \in J \mid xJ^n = (0)\}$; W is an ideal of R. If $W \supset J^n$ then $J^n J^n = (0)$ which would yield that $(0) = J^{2n} = J^n$ the desired outcome.

Suppose that $W \not\supset J^n$. Therefore in $\overline{R} = R/W$, $\overline{J}^n \neq (0)$. If $\bar{x}\overline{J}^n = (0)$ then $xJ^n \subset W$ hence $(0) = xJ^n J^n = xJ^{2n} = xJ^n$ placing x in W and so implying that $\bar{x} = 0$. That is, $\bar{x}\overline{J}^n = (0)$ forces $\bar{x} = 0$.

Since $\overline{J}^n \neq (0)$ it contains a minimal right ideal $\bar{\rho} \neq (0)$ of $\overline{R}$. But in that event $\bar{\rho}$ is an irreducible $\overline{R}$-module hence is annihilated by $J(\overline{R})$. Since $\overline{J}^n \subset J(\overline{R})$ we get $\bar{\rho}\overline{J}^n = (0)$. As we have seen above this forces a contradiction $\bar{\rho} = (0)$. The theorem is now proved.

The theorem has as a consequence a well-known result first proved by Hopkins.

COROLLARY. *If R is Artinian then any nil ideal (right left, or two-sided) of R is nilpotent.*

Proof. In view of Lemma 1.2.2 a nil one-sided ideal is in $J(R)$; $J(R)$ being nilpotent we infer the corollary.

The corollary is capable of a simple, independent

proof as follows: let ρ be a nil right ideal of R. The descending chain $\rho \supset \rho^2 \supset \cdots \supset \rho^n \supset \cdots$ stops, so that for some n, $B = \rho^n = \rho^{n+1} = \cdots$. Therefore $B^2 = B$. If $B \neq (0)$, from $B^2 = B \neq (0)$ there is a $b \in B$ such that $bB \neq (0)$. Pick $b_0 \in B$ so that $b_0 B \neq (0)$ is minimal of the form $bB \neq (0)$. Since $b_0 B = b_0 B^2 \neq (0)$ there is a $c \in B$ with $b_0 c B \neq (0)$. Now $(0) \neq b_0 c B \subset b_0 B$ yields, from the minimality of $b_0 B$, that $b_0 c B = b_0 B$. Therefore there is an $x \in B$ such that $b_0 c = b_0 c x$; this gives $0 \neq b_0 c = b_0 c x = b_0 c x^2 = \cdots = b_0 c x^n = 0$ for a large enough n (since $x \in B$ must be nilpotent). This contradiction proves that $B = (0)$ and, so, that ρ is nilpotent.

Although the results we are about to present are coming a little out of sequence this seems an appropriate place to develop them.

First we make a simple observation. Let R be any ring and suppose that $\rho \neq (0)$ is a nilpotent right ideal of R. If $R\rho = (0)$ then $R\rho$ is certainly contained in ρ whence ρ is a two-sided ideal of R. If $R\rho \neq (0)$ it is a two-sided ideal of R; moreover, if $\rho^m = (0)$ then $(R\rho)^m = R\rho R \cdots R\rho = R(\rho R)(\rho R) \cdots (\rho R)\rho \subset \rho^m = (0)$. In summary, what we have shown is that if R contains a nonzero nilpotent right ideal it contains a nonzero nilpotent two-sided ideal. The validity of the analogous result for nil right ideals is an open question called the *Köthe conjecture*. It asks whether a ring which has a nonzero nil one-sided ideal must have a nonzero nil two-sided ideal.

DEFINITION. *An element $e \neq 0$ in R is an idempotent if $e^2 = e$.*

LEMMA 1.3.1. *Let R be a ring having no nonzero nilpotent ideals. Suppose that $\rho \neq (0)$ is a minimal right ideal of R. Then $\rho = eR$ for some idempotent e in R.*

Proof. As R has no nilpotent ideals, by the remark

made above $\rho^2 \neq (0)$ hence there is an $x \in \rho$ such that $x\rho \neq (0)$. However $x\rho \subset \rho$ is a right ideal of R which, by virtue of the minimality of ρ, yields that $x\rho = \rho$. Therefore there is an element $e \in \rho$ satisfying $xe = x$. Since $xe^2 = xe$ we get that $x(e^2 - e) = 0$. Let $\rho_0 = \{a \in \rho \mid xa = 0\}$; ρ_0 is a right ideal of R, is contained in ρ and is not ρ since $x\rho \neq (0)$. We must have, therefore, that $\rho_0 = (0)$; $e^2 - e$ being in ρ_0 yields that $e^2 = e$. Since $xe = x \neq 0$ we have that $e \neq 0$. Now $eR \subset \rho$ is a right ideal of R and contains $e^2 = e \neq 0$ so that $eR \neq (0)$. The net result of this is that $eR = \rho$. This is precisely the assertion of the lemma.

We have seen that in an Artinian ring a right ideal consisting of nilpotent elements must itself be nilpotent. What about the contrary case in which a right ideal has a nonnilpotent element? Our aim is to show that it must then contain a nonzero idempotent. To establish this we need the next result which by itself has some independent interest.

LEMMA 1.3.2. *Let R be a ring and suppose that for some $a \in R$, $a^2 - a$ is nilpotent. Then either a is nilpotent or, for some polynomial $q(x)$ with integer coefficients $e = aq(a)$ is a nonzero idempotent.*

Proof. Suppose that $(a^2 - a)^k = 0$; expanding this we obtain $a^k = a^{k+1}p(a)$ where $p(x)$ is a polynomial having integer coefficients. We feed the value of a^k into the right hand side to get $a^k = a^k a p(a) = a^{k+2} p(a)^2$. Continuing we get $a^k = a^{2k} p(a)^k$. If $a^k \neq 0$ then $e = a^k p(a)^k \neq 0$ and $e^2 = a^{2k} p(a)^{2k} = a^k p(a)^k = e$.

We now are able to prove the previously mentioned result.

THEOREM 1.3.2. *If R is an Artinian ring and $\rho \neq (0)$ is a nonnilpotent right ideal of R then ρ contains a nonzero idempotent.*

Proof. Since ρ is not nilpotent, by Theorem 1.3.1 it is not contained in $J(R)$. Let $\overline{R} = R/J(R)$; as a semisimple ring $\overline{R}$ has no nonzero nilpotent ideals. Let $\bar{\rho}$ be the image of ρ in $\overline{R}$. Since $\bar{\rho} \neq (0)$ it contains a minimal right ideal $\bar{\rho}_0$ of $\overline{R}$. Now by Lemma 1.3.1 $\bar{\rho}_0$ has an idempotent $\bar{e} \neq 0$. Let $a \in \rho$ map onto $\bar{e}$; hence $a^2 - a$ maps onto 0 so is in $J(R)$ in consequence of which it must be nilpotent. Because $\bar{a}^k = \bar{e}^k = \bar{e} \neq 0$ a is not nilpotent. By the previous lemma we obtain that for some polynomial with integer coefficients $e = aq(a)$ is a nonzero idempotent. Since $a \in \rho$, e must also be in ρ. This completes the proof.

Since we have been discussing idempotents it is not too unnatural to investigate the nature of the ring eRe where e is an idempotent in R. This ring is closely related to R. We calculate it now in a specific instance. Let $R = A_n$ where A is any ring with unit element and let

$$
e = \begin{pmatrix} 1 & & & & & \\ & 1 & & & & 0 \\ & & \cdot & & & \\ & & & 1 & & \\ & & & & 0 & \\ & & & & & \cdot \\ 0 & & & & & 0 \end{pmatrix}
$$

be an idempotent of rank r. The outcome of calculating eRe is that

$$
eRe = \left\{ \left(\begin{array}{c|c} a & 0 \\ \hline 0 & 0 \end{array} \right) \;\middle|\; a \in A_r \right\},
$$

that is $eRe \approx A_r$, a matrix ring of smaller size.

THEOREM 1.3.3. *Let R be any ring and let e be an idempotent in R. Then $J(eRe) = eJ(R)e$.*

Proof. Let M be an irreducible R-module. We assert that either $Me = (0)$ or Me is an irreducible eRe-module. Suppose that $Me \neq (0)$ and that $me \neq 0 \in Me$. Now $(me)(eRe) = meRe$; since M is an irreducible R-module and $me \neq 0$, $meR = M$ hence $meRe = Me$. In this way Me turns out to be an irreducible eRe-module. Thus $MeJ(eRe = (0)$; since e is a unit for eRe hence $eJ(eRe) = J(eRe)$. In other words if $Me \neq (0)$ then $MJ(eRe) = (0)$. If $Me = (0)$ then certainly $(0) = MeJ(eRe) = MJ(eRe)$. All in all, $J(eRe)$ annihilates all irreducible R-modules M, hence $J(eRe) \subset J(R)$. This yields $J(eRe) = eJ(eRe)e \subset eJ(r)e$.

On the other hand, if $a \in eJ(R)e$ then, as an element of $J(R)$ it has a right and left quasi-inverse a'. From $a + a + aa' = 0$, on multiplication from left and right by e and by use of $eae = a$ we get $a + ea'e + aea'e = 0$. Since the quasi-inverse of a is unique we end up with $a' = ea'e$. We have shown that every element in $eJ(R)e$ is quasi-regular in eRe. Moreover $eJ(R)e$ is an ideal of eRe; as a quasi-regular ideal of eRe it must be in $J(eRe)$. Having established $J(eRe) \subset eJ(R)e \subset J(eRe)$ we reach the desired conclusion $J(eRe) = eJ(R)e$.

We continue with our examination of the interplay between R and eRe. Is there anything special we can say about eRe when e itself is something special? One specialization we can make on e is that the right ideal eR be rather restricted. This is the direction taken in

THEOREM 1.3.4. *Let R be a ring having no nonzero nilpotent ideals and suppose that $e \neq 0$ is an idempotent in R. Then eR is a minimal right ideal of R if and only if eRe is a division ring.*

Proof. First let us suppose that $\rho = eR$ is a minimal

right ideal of R. If $eae \neq 0 \in eRe$ then $(0) \neq eaeR \subset eR$ yields that $eaeR = eR$. Hence there is a $y \in R$ such that $eaey = e$, $eaeye = e^2 = e$ which is to say, $(eae)(eye) = e$. Thus eRe is a division ring with unit element e.

If, on the other hand, eRe is a division ring we claim that $\rho = eR$ is a minimal right ideal of R. For, if $(0) \neq \rho_0 \subset \rho$ is a right ideal of R then $\rho_0 e \neq (0)$ otherwise $\rho_0{}^2 \subset \rho_0 \rho = \rho_0 eR = (0)$, which is obviously impossible in a ring without nilpotent ideals. Let $a = eae \neq 0$ be in eRe; since $a \in eR$ and $ea = a$ we have $0 \neq ae = eae \in \rho_0$; being a nonzero element in the division ring eRe, for some $exe \in eRe$, $eaexe = e$. However $e = eaeexe \in \rho_0$; this forces $eR \subset \rho_0 \subset eR = \rho$. Hence $\rho_0 = \rho$ and ρ is indeed minimal.

COROLLARY. *If R has no nonzero nilpotent ideals and if e is an idempotent in R then $e R$ is a minimal right ideal of R if and only if Re is a minimal left ideal of R.*

Proof. Clearly interchanging left and right in the previous theorem we get that Re is a minimal left ideal of R if and only if eRe is a division ring, hence if and only if eR is a minimal right ideal of R.

4. Semisimple Artinian rings.

Up to this point we have been studying the radical of a ring. We now change focus and study rings whose radicals are as trivial as possible, namely, are (0). We postpone the discussion of general such rings until the next chapter contenting ourselves for the moment with a dissection of semisimple Artinian rings.

However, before going on to this study we should like to establish that semisimple rings do exist in nature, so to speak. In other words, we want to show that some naturally defined class of rings consists of semisimple Artinian rings. The result we obtain is a highly important classical theorem of Maschke.

Let F be a field and let G be a finite group of order $o(G)$. By the *group algebra* of G over F, which we write as $F(G)$, we mean $\{\sum \alpha_i g_i \,|\, \alpha_i \in F,\ g_i \in G\}$ where the group elements are considered to be linearly independent over F, where addition is in the natural way and where multiplication is by use of the distributive laws and the calculation $g_i g_j$ according to the product in G.

Maschke's Theorem then reads as

THEOREM 1.4.1. *Let G be a finite group of order $o(G)$ and let F be a field of characteristic 0 or of characteristic p where $p \nmid o(G)$. Then $F(G)$ is semisimple.*

Proof. For $a \in F(G)$ we define the mapping $T_a \colon F(G) \to F(G)$ by $xT_a = xa$ for all $x \in F(G)$. This is called the *right regular representation* of $F(G)$.

As is immediately verified, T_a is an F-linear transformation on $F(G)$. Moreover the mapping $\Psi \colon a \to T_a$ is an algebra isomorphism of $F(G)$ into the algebra of linear transformations on $F(G)$ considered as a vector space over F. We represent T_a as a matrix using the group elements as a basis for $F(G)$ over F. We note:

1. if $g \neq 1 \in G$ then the trace of T_g, tr T_g, is 0. (tr $T_g = 0$)
2. tr $T_1 = o(G)$.

Let J be the radical of $F(G)$; since $F(G)$ is finite-dimensional over F, so Artinian as an algebra, J is nilpotent by Theorem 1.3.1. Suppose that $0 \neq x = \alpha_1 g_1 \cdots + \alpha_n g_n \in J$. Since J is an ideal of $F(G)$ by multiplying by the appropriate g_i^{-1} we may assume that $x = \alpha_1 + \alpha_2 g_2 + \cdots + \alpha_n g_n$ with $\alpha_1 \neq 0$.

However, since $x \in J$ it is nilpotent hence T_x is a nilpotent linear transformation. As such, tr $T_x = 0$. We calculate tr T_x. From $x = \alpha_1 + \alpha_2 g_2 + \cdots + \alpha_n g_n$ we have that $T_x = \alpha_1 T_1 + \alpha_2 T_{g_2} + \cdots + \alpha_n T_{g_n}$ and so

$$0 = \text{tr } T_x = \alpha_1 \text{ tr } T_1 + \alpha_2 \text{ tr } T_{g_2} + \cdots + \alpha_n \text{ tr } T_{g_n}$$
$$= \alpha_1 o(G)$$

by the remark made above. Since $\alpha_1 \neq 0$ and $\alpha_1 o(G) = 0$ we conclude that $o(G) = 0$ in F. In characteristic 0 this is nonsense; in characteristic $p \neq 0$ this implies that $p \mid o(G)$ contrary to assumption. The theorem is thereby proved. We point out that $F(G)$ definitely is not semisimple when the characteristic of F divides $o(G)$. For let $a = \sum_{g \in G} g$; since the group elements are linearly independent $a \neq 0$. For any $x \in G$, $ax = a = xa$, placing a thereby in the center of $F(G)$. Now $a^2 = a \sum g = o(G)a = 0$ since $p \mid o(G)$. Thus the ideal $F(G)a$ satisfies $(F(G)a)^2 = F(G)aF(G)a = F(G)^2 a^2 = (0)$. Having produced a nonzero nilpotent ideal in $F(G)$ we know that $F(G)$ cannot be semisimple.

In the special case in which F is the field of complex numbers, or any subfield thereof, a different argument can be given which may be of interest to the reader. We merely give the main outline and steps of the proof leaving some of the details to the reader.

Let F be the field of complex numbers and let $\Gamma = F(G)$ for G a finite group. For $\alpha \in F$ let $\bar{a}$ be its complex conjugate. We define a map $*$ on Γ into Γ. If $x = \sum \alpha_i g_i \in \Gamma$ define $x^* = \sum \bar{a}_i g_i^{-1}$. The following are clear:

1. $x^{**} = x$
2. $(\alpha x + \beta y)^* = \bar{a}x^* + \bar{\beta}y^*$
3. $(xy)^* = y^*x^*$ for all $x, y \in \Gamma$, $\alpha, \beta \in F$.

If $x = \sum \alpha_i g_i$ then $xx^* = \sum |\alpha_i|^2 + \sum_{g_k \neq 1} \beta_k g_k$ hence $xx^* = 0$ forces $\sum |\alpha_i|^2 = 0$ and so each $\alpha_i = 0$, that is $xx^* = 0$ implies that $x = 0$. Let u be in J; then $uu^* = v$ is in J. Now $v^* = v$ and as an element of J v must be nilpotent. From the fact that $xx^* = 0$ implies $x = 0$ we deduce that $v = 0$ hence $uu^* = 0$ leading to $u = 0$. Thus $J = (0)$ and so J is semisimple.

The argument just given can be adapted to the case when G is any group, not necessarily finite, to show that $F(G)$ is semisimple in case F is the field of complex numbers. The means to this is to realize $F(G)$ as a ring of operators on a Hilbert space. This observation was first made by Rickart. Amitsur and we independently showed that over an uncountable field of characteristic 0 $F(G)$ is semisimple for any group. To settle the question of the semisimplicity of $F(G)$ for any group G over a field F of characteristic 0 one can show it is enough to settle it for the special case in which F is the field of rational numbers.

This seems an opportune moment to mention some other open questions about the structure of $F(G)$ for G an arbitrary group. We begin with the one just mentioned.

1. If F is of characteristic 0 is $F(G)$ semisimple?

2. If G has no elements of finite order is $F(G)$ semisimple for any field F?

3. If G has no elements of finite order is $F(G)$ free of zero divisors?

4. If $F(G)$ is algebraic—that is, every element in $F(G)$ satisfies a polynomial over F—is G locally finite—that is, does every finite subset of G generate a finite subgroup? If F is of characteristic 0 we showed this to be the case. For characteristic $p \neq 0$ the problem remains open.

5. If every element in $F(G)$ is of the form $\alpha + n$, $\alpha \in F$, n nilpotent, is G locally finite? This is a special case of (4). It is easy to show that F must be of characteristic $p \neq 0$ and that every element in G is of finite order, the order being a power of p.

There is recent literature on $F(G)$ for a general group G by Amitsur, M. Auslander, McLaughlin, Passman and Villamayor.

We now return to our study of semisimple Artinian

rings. Lemma 1.3.1 revealed that a minimal right ideal in such a ring is generated by an idempotent. The condition that the right ideal be minimal is not essential for the conclusion. This is

THEOREM 1.4.2. *Let R be a semisimple Artinian ring and let $\rho \neq (0)$ be a right ideal of R. Then $\rho = eR$ for some idempotent e in R.*

Proof. As a nonzero right ideal in a semisimple ring ρ can not be nilpotent, hence by Theorem 1.3.2 it must have a non-zero idempotent. If e is an idempotent in ρ let $A(e) = \{x \in \rho \,|\, ex = 0\}$. The set of right ideals $\{A(e) \,|\, e^2 = e \neq 0 \in \rho\}$ is a nonempty set so has a minimal element $A(e_0)$.

If $A(e_0) = (0)$ then, since for any $x \in \rho$ $e_0(x - e_0 x) = 0$, $x - e_0 x \in A(e_0) = (0)$ resulting in $x = e_0 x$ for all $x \in \rho$. But this in its turn would imply that $\rho = e_0 \rho \subset e_0 R \subset \rho$, the last relation coming from $e_0 \in \rho$. This is the contention of the theorem.

Suppose then that $A(e_0) \neq (0)$; we will show that this is not possible. As a nonzero right ideal of R, $A(e_0)$ must have an idempotent e_1. By the definition of $A(e_0)$ e_1 is in ρ and $e_0 e_1 = 0$. Consider $e^* = e_0 + e_1 - e_1 e_0$. It is in ρ and a simple computation reveals that it is an idempotent. Moreover $e^* e_1 = (e_0 + e_1 - e_1 e_0)e_1 = e_1 \neq 0$, hence, in particular, $e^* \neq 0$. Now, if $e^* x = 0$ then $(e_0 + e_1 - e_1 e_0)x = 0$ hence $e_0(e_0 + e_1 - e_1 e_0)x = 0$; this yields that $e_0 x = 0$. In other words, $A(e^*) \subset A(e_0)$; since $e_1 \in A(e_0)$ and $e_1 \not\in A(e^*)$ we have that $A(e^*) \neq A(e_0)$. Since $A(e_0)$ is minimal and $A(e^*)$ is properly contained in $A(e_0)$ we have reached a contradiction. Thus the possibility $A(e_0) \neq (0)$ cannot arise and so the theorem is proved.

The theorem has two very interesting corollaries, each of them theorems in their own right.

COROLLARY 1. *If R is a semisimple Artinian ring and A is an ideal of R then $A = eR = Re$ where e is an idempotent in the center of R.*

Proof. Since A is a right ideal of R, by the theorem $A = eR$ for some idempotent e. Let $B = \{x - xe \mid x \in A\}$; clearly since $ex = x$ for all $x \in A$ and $Be = (0)$ so $BA = BeA = (0)$. However, as A is also a left ideal of R, B must be a left ideal of R. Moreover $B^2 \subset BA = (0)$; the absence of nilpotent ideals in R implies that $B = (0)$, that is, $x = xe$ for all $x \in A$. From this we immediately have that $A = Re$. In this way e is both a left and right unit element for A.

Now we show that e is in the center of R. Let $y \in R$; then $ye \in A$ hence $ye = e(ye)$ because e is a left unit for A. In addition, $ey \in A$ hence $ey = (ey)e$ in virtue of the fact that e is a right unit for A. Combining the pieces we have shown that $ye = eye = ey$ for all $y \in R$, that is, e is in the center of R.

The next result which itself is a corollary to Corollary 1 has also a great interest.

COROLLARY 2. *A semisimple Artinian ring has a two-sided unit element.*

Proof. R is an ideal of R so the result comes directly from Corollary 1.

This last fact asserts that semisimplicity forces the existence of a unit element in an Artinian ring, something not *a priori* obvious. We digress to consider a related problem.

A ring R is said to be *torsion-free* if $mx = 0$, with m an integer, x in R, only if $m = 0$ or $x = 0$. We propose to establish the existence of a right unit element in any torsion-free right Artinian ring, be it semisimple or not. First to some subsidiary results.

LEMMA 1.4.1. *Let R be a torsion-free right Artinian ring. Then,*

1. *if $xR = (0)$ for $x \in R$ then $x = 0$*

2. *if $m \neq 0$ is an integer then $mR = R$ (and so R is an algebra over the rationals)*

3. *if J is the radical of R then R/J^k is torsion free for all integers $k \geq 1$.*

Proof. If $xR = (0)$ then $I_n = \left\{ k2^n x \mid k \text{ any integer} \right\}$ is a right ideal of R. Moreover, since R is torsion free the chain $I_0 \supset I_1 \supset \cdots \supset I_n \supset \cdots$ is a properly descending chain, unless $x = 0$.

If $m \neq 0$ is an integer then $R \supset mR \supset m^2R \supset \cdots m^kR \supset m^{k+1}R \supset \cdots$ is a descending chain of ideals so must terminate. Thus for some k, $m^kR = m^{k+1}R$ which, in the presence of the torsion-freeness of R, implies that $R = mR$. In particular, given $x \in R$, $x = my$ for some y; y is unique for $x = my = mz$ yields $m(y - z) = 0$ and so $y = z$. Write y as $(1/m)x$; it is easy to verify that this renders R into an algebra over the rational field Q.

As an algebra over Q the radical J of R, as we have seen earlier, is an algebra ideal as are all the J^k, hence R/J^k is an algebra over Q, thus is torsion free.

Of some independent interest is

LEMMA 1.4.2. *Let R be a ring in which $xR = (0)$ implies that $x = 0$. Suppose that A, B are ideals of R such that:*

1. *R/A has a right unit element.*

2. *R/B has a left unit element.*

3. *$AB = (0)$.*

Then R has a right unit element.

Proof. Let e in R map on the right unit of R/A and let f in R map on the left unit of R/B. Hence for x, $y \in R$, $x - xe \in A$, $y - fy \in B$. Therefore $(x - xe)(y - fy) \in AB = (0)$. This says that $((x - xe)f - (x - xe))R = (0)$

for all $x \in R$. By assumption we get $(x - xe)f - (x - xe) = 0$ and so $x = x(e + f - ef)$ for all $x \in R$. The element $e + f - ef$ is the desired right unit.

We have all the required information to prove the previously mentioned theorem, namely

THEOREM 1.4.3. *A torsion-free right Artinian ring has a right unit element.*

Proof. Let J be the radical of R; as we have seen, J must be nilpotent. The *index of nilpotence* of J is that integer k such that $J^k = (0)$, $J^{k-1} \neq (0)$. We proceed by induction on the index of nilpotence of J which we write as k.

If $k = 1$ then $J = (0)$ so that R is semisimple, hence, by Corollary 2 to Theorem 1.4.2, R has a unit element.

Suppose that $k > 1$ and let $\overline{R} = R/J^{k-1}$. In $\overline{R}$ the radical is J/J^{k-1} which has index of nilpotence $k - 1$. By induction $\overline{R}$ has a right unit element. Now R/J being a semisimple Artinian ring has a two-sided unit element, therefore a left unit element. In addition $J^{k-1}J = (0)$ so the previous lemma applies with $A = J^{k-1}$, $B = J$. In this way R has been shown to have a right unit element.

We return to semisimple Artinian rings. Let R be such a ring and let $A \neq (0)$ be an ideal of R. By Corollary 1 to Theorem 1.4.2, $A = eR = Re$ where e is an idempotent in the center of R. By Corollary 2 to this same theorem R has a unit element 1.

Given $x \in R$ then $x = xe + x(1 - e)$ hence $R = Re + R(1 - e)$. This is the *Peirce decomposition* of R relative to e. Since $1 - e$ is also in the center of R, $R(1 - e)$ is an ideal of R. Moreover, $Re \cap R(1 - e) = (0)$ for if x is in this intersection, as an element of Re, $x = xe$ and as an element of $R(1 - e)$, $xe = 0$. Hence R is the direct sum of A and $R(1 - e)$. In particular A is isomorphic as a ring to $R/R(1 - e)$; as a homomorphic image of the Artinian ring R, A turns out to be Artinian. On the other hand, by the corollary to Theorem 1.2.5 A is semisimple. We have proved

LEMMA 1.4.3. *An ideal of a semisimple Artinian ring is a semisimple Artinian ring.*

DEFINITION. *A ring R is said to be simple if $R^2 \neq (0)$ and R has no ideals other than (0) and R.*

The condition $R^2 \neq (0)$ imposed in the last definition is to avoid the trivial possibility that R is an additive group with p elements, p a prime, in which the product of any two elements is 0. If R has a unit element it is trivial to show that simplicity implies semisimplicity. There are examples, due to Sasiada, of simple rings which are their own radicals. The question about the existence of simple radical rings had been open for some time when it was settled by Sasiada. We leave to the reader to show that a simple Artinian ring must be semisimple.

We give a nice example of a simple ring which has no zero divisors and yet is not a division ring. Let F be a field of characteristic 0 and let R be the ring of all polynomials $\sum \alpha_{ij} x^i y^j$, $\alpha_{ij} \in F$, with equality and addition as usual and with F considered in the center under multiplication and where, in addition, we multiply according to the commutation rule $xy - yx = 1$. We leave the verification of the cited properties of R to the reader.

We exploit the argument used in proving Lemma 1.4.3 further. Let $A \neq (0)$ be a minimal ideal in the semisimple Artinian ring R. We claim that A is a simple ring. As an ideal in a semisimple ring A certainly satisfies $A^2 \neq (0)$. If $B \neq (0)$ is an ideal of A then ABA is an ideal of R and is contained in B. Since A has a left unit, $AB \neq (0)$; as AB is a nonzero left ideal of R it is not nilpotent, hence in particular, $ABA \neq (0)$. By the minimality of A we arrive at $B \supset ABA = A$ yielding $B = A$. Thus A is a simple Artinian ring. As we saw in the proof of Lemma 1.4.3, $R = A \oplus T_0$ where T_0 is an ideal of R, so is semi-simple Artinian. Pick a minimal ideal A_1 of R lying in T_0. We now know that A_1 is simple and Artinian and that $T_0 = A_1 \oplus T_1$. Continuing we get ideals $A = A_0$, $A_1, \cdots, A_k, \cdots$ of R, all simple and Artinian and such that the sums $A_0 + \cdots + A_k$ are all direct.

We claim that for some k, $R = A_0 \oplus \cdots \oplus A_k$; if not, let $R_0 = A_0 \oplus \cdots \oplus A_k \oplus \cdots$, $R_1 = A_1 \oplus \cdots \oplus A_k \oplus \cdots$, $\cdots$, $R_m = A_m \oplus \cdots \oplus A_k \oplus \cdots$. This is a descending chain of ideals so it must terminate. But when it terminates we see from our construction that $R = A_0 \oplus \cdots \oplus A_k$. We have proved a famous theorem due to Wedderburn.

THEOREM 1.4.4. *A semisimple Artinian ring is the direct sum of a finite number of simple Artinian rings.*

The structure of semisimple Artinian rings will be completely determined once we characterize the simple ones. This we do in the next chapter in another famous theorem of Wedderburn.

In a theorem such as Theorem 1.4.4 it is natural to wonder about the uniqueness of the decomposition. This is answered for us in the next lemma.

LEMMA 1.4.4. *If R is a semisimple Artinian ring and $R = A_1 \oplus \cdots \oplus A_k$ where the A_i are simple then the A_i account for all the minimal ideals of R.*

Proof. Let $B \neq (0)$ be a minimal ideal of R. Since R has a unit element $RB \neq (0)$. But $RB = A_1B \oplus \cdots \oplus A_kB$ therefore, for some i, $A_iB \neq (0)$. Now A_iB is an ideal of R and $A_iB \subset A_i$ hence, by the minimality of A_i we get $A_iB = A_i$. Since B is minimal and $A_iB \subset B$ we get

$A_i B = B$, yielding as a consequence $A_i = A_i B = B$. B has been shown to be one of the A_i; this proves the lemma.

In Theorem 1.3.1 and its corollary we saw that in the presence of the descending chain condition on right ideals a nil right ideal is rendered nilpotent. We conclude the chapter by proving a companion piece to this, namely, that in the presence of the ascending chain condition on right ideals a nil right ideal must be nilpotent. This result is a special case of a theorem due to Levitzki; it has been extensively generalized, for instance in a paper by Herstein and Small.

DEFINITION. *A ring is said to be right Noetherian if any nonempty set of right ideals has an element maximal in it.*

It is worth remarking that this is equivalent to the fact that any ascending chain of right ideals must become stationary. One should also remark that one can define left Noetherian rings. Examples exist of rings which are right but not left Noetherian. We give two such here.

1. Let Q be the field of rational numbers and let $Q(x)$ be the field of rational functions over Q in the indeterminate x. We define a monomorphism $\alpha: Q(x) \rightarrow Q(x)$ by $\alpha(r(x)) = r(x^2)$. Let $R = \left\{ \sum y^i r_i(x) \mid r_i(x) \in Q(x) \right\}$ be the right polynomials in y over $Q(x)$, where equality and addition are as usual but where we multiply by means of the rule $r(x)y = y\alpha(r(x))$. It is easy to show that every right ideal of R is principal, and so R is right Noetherian, but that R is not left Noetherian.

2. Let A be a commutative Noetherian integral domain (for instance, the integers) and let F be its field of quotients. Let

$$R = \left\{ \begin{pmatrix} a & \alpha \\ 0 & \beta \end{pmatrix} \middle| a \in A,\, \alpha, \beta \in F \right\}.$$

It is easy to show that R is right Noetherian but is not left Noetherian. This simple family of examples is due to Lance Small.

Herstein used this last example to construct a counterexample to a conjecture in ring theory. Let A be the ring of all rational numbers whose denominators are odd. We leave to the reader to show that $J(A)$ consists of all rationals having odd denominator and even numerator. The field of quotients of A and of $J(A)$ is Q, the field of rational numbers. Let R be as above, that is,

$$R = \left\{ \begin{pmatrix} a & \alpha \\ 0 & \beta \end{pmatrix} \middle| a \in A, \alpha, \beta \in F \right\}.$$

We leave to the reader to show that

$$J(R) \supset \left\{ \begin{pmatrix} J(A) & Q \\ 0 & 0 \end{pmatrix} \right\} \quad \text{and so} \quad J(R)^2 \supset \left\{ \begin{pmatrix} 0 & Q \\ 0 & 0 \end{pmatrix} \right\}$$

and in fact

$$J(R)^n \supset \left\{ \begin{pmatrix} 0 & Q \\ 0 & 0 \end{pmatrix} \right\}.$$

Thus

$$\bigcap_n J(R)^n \supset \left\{ \begin{pmatrix} 0 & Q \\ 0 & 0 \end{pmatrix} \right\}$$

and so is not (0). The conjecture was that it was (0). It would be interesting to prove or disprove this conjecture if R is assumed to be any ring which is both left and right Noetherian.

We now prove Levitzki's Theorem; the proof used is due to Utumi.

THEOREM 1.4.5. *Let A be a nil one-sided ideal in a right Noetherian ring R. Then A is nilpotent.*

Proof. Since R is right Noetherian it has a maximal nilpotent ideal N. Our aim is to show that $A \subset N$. If not, by passing to $\overline{R} = R/N$ we reach the following situation; $\overline{R}$ is a right Noetherian ring which has no nonzero nilpotent ideals and $\overline{A} \neq (0)$ is a nil one-sided ideal of $\overline{R}$. We wish to show that this is impossible. In other words, we may assume without loss of generality, that R has no nilpotent ideals but has a nil one-sided ideal $A \neq (0)$.

If $a \neq 0 \in A$ then $U = Ra$ is a nil left ideal of R, for if A is a left ideal of R then since $U \subset A$ it, too, would be nil. If, on the other hand, A is a right ideal and $u = za \in U$ then $u^n = x(ax)^{n-1}a$ so is 0 for large enough n since $ax \in A$.

If $u \in U$ let $r(u) = \{x \in R \mid ux = 0\}$; $r(u)$ is a nonzero right ideal of R. R being Noetherian there is a $u_0 \neq 0$ in U with $r(u_0)$ maximal. For any $x \in R$ clearly $r(xu_0) \supset r(u_0)$ hence, if $xu_0 \neq 0$ since it is in U we get $r(xu_0) = r(u_0)$ from the maximality of $r(u_0)$. Let $y \in R$; then $(yu_0)^k = 0$, $(yu_0)^{k-1} \neq 0$ for some k. Because $(yu_0)^{k-1}$ is of the form xu_0 we have $r((yu_0)^{k-1}) = r(u_0)$. But yu_0 is in $r((yu_0)^{k-1})$ so is in $r(u_0)$; that is $u_0(yu_0) = 0$ for all $y \in R$. This says that u_0R is a nilpotent right ideal of R hence is (0). But then $\{t \in R \mid tR = (0)\}$ is a nonzero nilpotent right ideal (containing u_0). With this contradiction the theorem is proved.

References

1. S. A. Amitsur, A general theory of radicals I, *Amer. J. Math.*, 74 (1952) 774–776.

2. ———, A general theory of radicals II, *Amer. J. Math.*, 76 (1954) 100–125.

3. ———, A general theory of radicals III, *Amer. J. Math.*, 76 (1954) 126–136.

4. S. A. Amitsur, Radicals of polynomial rings, *Canad. J. Math.*, 8 (1956) 355–361.

5. ———, Algebras over infinite fields, *Proc. Amer. Math. Soc.*, 7 (1956) 35–48.

6. ———, Semi-simplicity of group algebras, *Michigan Math. J.*, 6 (1959) 251–253.

7. E. Artin, C. Nesbitt and R. Thrall, *Rings with minimum conditions*, University of Michigan, 1944.

8. M. Auslander, On regular group rings, *Proc. Amer. Math. Soc.*, 8 (1957) 658–664.

9. I. N. Herstein, *Theory of rings*, University of Chicago, Math. Notes, 1961.

10. ———, A theorem of Levitzki, *Proc. Amer. Math. Soc.*, 13 (1962) 213–214.

11. ———, A counter-example in Noetherian rings, *Proc. Nat. Acad. Sci.* U.S.A., 54 (1965) 1036–1037.

12. I. N. Herstein and Lance Small, Nil rings satisfying certain chain conditions, *Canad. J. Math.*, 16 (1964) 771–776.

13. Nathan Jacobson, *Structure of rings*, Amer. Math. Soc. Colloq. Publ., 37 (1964)

14. ———, Radical and semi-simplicity for arbitrary rings, *Amer. J. Math.*, 67 (1945) 300–320.

15. A. Kurosh, Radicals of rings and algebras, *Mat Sb.*, 33 (1945) 13–26.

16. J. Levitzki, On multiplicative systems, *Compositio Math.*, 8 (1950) 76–80.

17. J. McLaughlin, A note on regular group rings, *Michigan Math. J.*, 5 (1958) 127–128.

18. P. S. Passman, Nil ideals in group rings, *Michigan Math. J.*, 9 (1962) 375–384.

19. E. Sasiada, Solution of the problem on the existence of a simple radical ring. To appear.

20. Lance Small, An example in Noetherian rings, *Proc. Nat. Acad. Sci.* U.S.A., 54 (1965) 1035.

21. Y. Utumi. A theorem of Levitzki, *Amer. Math. Monthly*, 70 (1963) 286.

22. O. Villamayor, On the semi-simplicity of group algebras, *Proc. Amer. Math. Soc.*, 9 (1958) 621–627.

CHAPTER 2

SEMISIMPLE RINGS

The aim in defining the radical was to concentrate the bothersome behavior of a ring in a piece of it such that when this piece was removed the resulting ring was well enough behaved to permit some delicate dissection. The guidelines we choose for this dissection are the beautiful theorems of Wedderburn for the case of Artinian rings.

The success of this scheme is capable of a somewhat objective measure in the results that eventually come forth. We now undertake a more minute study of semisimple rings. In the material developed we shall provide ourselves with an assortment of instruments to attack general ring-theoretic questions, questions in whose hypotheses the derived concepts of radical, semisimplicity and the like play no role, but in whose solutions they enter intimately.

As often as possible we shall specialize the theorems obtained to study their implications in the classical case, obtaining thereby many well-known results.

1. The density theorem. We begin with a basic concept in the structure theory of rings. The special rings we introduce play the analagous role for general semisimple rings as that played by simple rings in the Artinian case.

DEFINITION. *A ring R is said to be a* primitive *ring if it has a faithful irreducible module.*

Such a ring should really be called right primitive for all modules used are right modules. We could similarly define left primitive rings. The two notions are different as has been shown recently by Bergman who constructed a ring which is right but not left primitive.

As we saw in Lemma 1.1.1, if M is an irreducible R-module and $A(M) = \{x \in R \mid Mx = (0)\}$ then $R/A(M)$ is primitive. In particular, if ρ is a maximal regular right ideal of R and if $M = R/\rho$ then $A(M) = (\rho: R)$ hence by the remark above, $R/(\rho: R)$ is a primitive ring. Since $(\rho: R)$ is the largest two-sided ideal of R lying in ρ we see that a ring R which has a maximal regular right ideal containing no ideal of R must be primitive, and conversely. In particular, a primitive ring must be semisimple for $J(R) = \bigcap(\rho: R)$ where ρ ranges over the maximal regular right ideals of R; hence if $(\rho: R) = (0)$ for one such ρ we have $J(R) = (0)$. Should R in addition be commutative then it must be a field, for R is then both semisimple and contains no nontrivial ideals. We summarize these remarks in

THEOREM 2.1.1. *A ring R is primitive if and only if there exists a maximal regular right ideal ρ in R such that $(\rho: R) = (0)$. R is therefore semisimple; if it is also commutative it must be a field.*

We pointed out earlier that examples exist of simple rings which are their own radicals. But it is trivial to show that a simple ring which is semisimple is primitive. Thus primitivity is a generalization, except for some pathological cases, of simplicity.

Let R be a primitive ring and let M be a faithful irreducible module for R. If $C(M) = \Delta$ is the commuting ring of R on M then by Schur's lemma (Theorem 1.1.1) Δ is a division ring. We can consider M as a right vector space over Δ where $m\alpha$, for $m \in M$, $\alpha \in \Delta$, is interpreted as the action of α as an element in $E(M)$ on m.

DEFINITION. *R is said to act* densely *on M (or R is said* *to be* dense *on* M) *if for every n and $v_1, \cdots, v_n$ in M which are linearly independent over Δ and any n elements $w_1, \cdots, w_n$ in M there is an element $r \in R$ such that $w_i = v_i r$ for $i = 1, 2, \cdots, n$.*

Note that in case M is finite dimensional over Δ and if R acts both faithfully and densely on M then it must be isomorphic to $\mathrm{Hom}_\Delta(M, M) = \Delta_n$, the ring of all $n \times n$ matrices over Δ, where $n = \dim_\Delta M$. Thus density is a generalization of the ring of all linear transformations. This can be pushed even further. If we give to M the discrete topology and to $\mathrm{Hom}_\Delta(M, M)$ the compact open topology then R acts densely on M if and only if R is dense in $\mathrm{Hom}_\Delta(M, M)$ in this topology.

The basic result from which the whole structure theory flows is the *Density Theorem* due to Jacobson and Chevalley.

THEOREM 2.1.2. (Density Theorem.) *Let R be a primitive ring and let M be a faithful irreducible R-module. If $\Delta = C(M)$ then R is a dense ring of linear transformations on M over Δ.*

Proof. We first note that to prove the theorem it suffices to show that given V, a finite dimensional subspace of M over Δ, and an $m \in M$, $m \notin V$ then we can find an $r \in R$ with $Vr = (0)$ but $mr \neq 0$.

To see this, suppose we can always find such an r. Then $mrR \neq (0)$ so, by the irreducibility of M, $mrR = M$. We can thus find an $s \in R$ with mrs arbitrary and $Vrs = (0)$. Given $v_1, \cdots, v_n \in M$ linearly independent over Δ and $w_1, \cdots, w_n$ in M let V_i be the linear span over Δ of the v_j with $j \neq i$. Since $v_i \notin V_i$ we can find a $t_i \in R$ with $v_i t_i = w_i$, $V_i t_i = (0)$. If $t = t_1 + \cdots + t_n$ we see that $v_i t = w_i$ for $i = 1, 2, \cdots, n$, thereby exhibiting the density of R on M.

We now propose to prove that given $V \subset M$ of finite dimension over Δ and $m \in M$, $m \notin V$ then there exists an $r \in R$ such that $Vr = (0)$ but $mr \neq (0)$. We proceed by induction on the dimension of V over Δ.

Now $V = V_0 + w\Delta$ where dim $V_0 = \dim(V) - 1$ and where $w \notin V_0$. By our induction, if $A(V_0) = \{x \in R \mid V_0 x = (0)\}$ then for $y \notin V_0$ there is an $r \in A(V_0)$ such that $yr \neq 0$. Otherwise put, if $mA(V_0) = (0)$ then $m \in V_0$.

$A(V_0)$ is a right ideal of R; since $w \notin V_0$, $wA(V_0) \neq (0)$, hence, as a submodule of M, $wA(V_0) = M$. Suppose that $m \in M$, $m \notin V$ enjoys the property that whenever $Vr = (0)$ then $mr = 0$. We want to show that this is not possible. We define $\tau: M \to M$ by: if $x \in M$ and so $x = wa$, $a \in A(V_0)$, then $x\tau = ma$. We claim that τ is well defined. For if $x = 0$ then $0 = x = wa$, hence, a annihilates both V_0 (being in $A(V_0)$) and w, therefore it annihilates all of V. By assumption $ma = 0$. This says that $x = x\tau = ma = 0$, in short, that τ is well defined.

Clearly $\tau \in E(M)$; moreover if $x = wa$ with $a \in A(V_0)$ then for any $r \in R$, since $ar \in A(V_0)$, $xr = (wa)r = w(ar)$ hence $(xr)\tau = m(ar) = (ma)r = (x\tau)r$. This puts τ in Δ. Hence for $a \in A(V_0)$ $ma = (wa)\tau = (w\tau)a$, that is $(m - w\tau)a = 0$ for all $a \in A(V_0)$. By our induction hypothesis $m - w\tau \in V_0$ hence $m \in V_0 + w\tau \subset V_0 + w\Delta = V$. With this contradiction the proof is complete.

With this result established the road is clear for deriving a series of theorems. But first we note that its converse is also true. In fact if V is a vector space over a division ring D and R is a ring of linear transformations which is *singly transitive*—that is, given $v \neq 0$ in V and $w \in V$ there is an $r \in R$ with $w = vr$—then R is primitive. For, as a ring of linear transformations on V, R has V as a faithful module. The transitivity of R on V implies that V is irreducible as an R-module. Having V as a faithful irreducible module R must be primitive. However, the commuting ring of R on V need not be D—it certainly contains D—and may in fact be larger. In the examples preceding Lemma 1.1.3 we saw instances of this phenomenon.

Suppose, however, that R is a *doubly transitive* ring of linear transformations on a vector space V over a division ring D. By this we mean, given v_1, $v_2 \in V$ linearly independent over D and w_1, $w_2 \in V$ then there is an $r \in R$ such that $v_1 r = w_1$, $v_2 r = w_2$. As we saw above, R is a primitive ring with V as faithful irreducible module. What is the commuting ring, Δ, of R on V? Certainly, as before $D \subset \Delta$; by Schur's lemma Δ is a division ring.

Suppose that $D \neq \Delta$. If $\tau \in \Delta$, $\tau \notin D$ and if $v \neq 0 \in V$ then v, $v\tau$ are linearly independent over D. If not, $v = v\tau\alpha$, $\alpha \in D$, hence $v(1 - \tau\alpha) = 0$ forcing $1 - \tau\alpha = 0$ and so $\tau = \alpha^{-1} \in D$. However, since v and $v\tau$ are linearly independent over D, by the double transitivity of the action of R on V, there is an $r \in R$ such that $vr = 0$ and $(v\tau)r = v \neq 0$. As $\tau \in \Delta$ this yields that $0 \neq v = (v\tau)r = (vr)\tau = 0$, which is nonsense. We have shown that $\Delta = D$. By the density theorem we then know that R is n-fold transitive on V over D for any n for which we can find n linearly independent elements in V over D. We summarize these remarks in

THEOREM 2.1.3. *If R is a doubly transitive ring of linear transformations on a vector space V over a division ring D then R is dense on V and the commuting ring of R on V is precisely D.*

The density theorem allows us to draw many conclusions about primitive rings and to relate them to matrix rings.

THEOREM 2.1.4. *Let R be a primitive ring. Then for some division ring Δ either R is isomorphic to Δ_n, the ring of all $n \times n$ matrices over Δ or, given any integer m there* *exists a subring S_m of R which maps homomorphically onto Δ_m.*

Proof. R acts as a dense ring of linear transformations on a vector space V over a division ring Δ. If V is finite dimensional over Δ then the density of R on V tells us that R is isomorphic to the ring of all Δ-linear transformations on V, that is, to Δ_n where $n = \dim_\Delta V$.

If V is not finite dimensional over Δ let $v_1, v_2, \cdots, v_m, \cdots$ be an infinite linearly independent set of elements of V. Let $V_m = v_1\Delta + \cdots + v_m\Delta$ and let $S_m = \{x \in R \mid V_m x \subset V_m\}$. The density theorem says that any Δ-linear transformation can be induced by an element of R. If $W_m = \{x \in S_m \mid V_m x = (0)\}$ then this translates into: S_m/W_m is isomorphic to Δ_m. This is the assertion of the theorem.

We extend a familiar concept from the theory of commutative rings to a noncommutative setting. The family of rings so defined contains all the primitive ones.

DEFINITION. *A ring R is said to be a prime ring if $aRb = (0)$, $a, b \in R$, implies that $a = 0$ or $b = 0$.*

The following equivalent formulations of primeness are easy to establish; we leave the proof to the reader.

LEMMA 2.1.1. *A ring R is a prime ring if and only if:*

1. *the right annihilator of a nonzero right ideal of R must be (0).*

2. *the left annihilator of a nonzero left ideal of R must be (0).*

3. *if A, B are ideals of R and $AB = (0)$ then either $A = (0)$ or $B = (0)$.*

As we shall see later a properly conditioned prime ring can be very tightly characterized. This is the content of a very beautiful theorem due to Goldie. In the meantime we show the primitive rings are always prime.

LEMMA 2.1.2. *A primitive ring is prime.*

Proof. Let $\rho \neq (0)$ be a right ideal of R, and suppose that $\rho a = (0)$. Since R is primitive it acts faithfully and irreducibly on an R-module M; because R is faithful on M, $M\rho \neq (0)$. Hence $M\rho = M$ follows from the irreducibility of M; this yields that $Ma = M\rho a = M(0) = (0)$ forcing $a = 0$. R has been shown to be prime.

Lemma 2.1.2 quickly yields that the center of a prime ring is an integral domain—it may be (0). For if Z is the center of the prime ring R and if $ab = 0$ with $a \in Z$, $b \neq 0$ in R then $(0) = Rab = aRb$; the primeness of R and $b \neq 0$ yields that $a = 0$. Thus

LEMMA 2.1.3. *A nonzero element in the center of a prime ring R is not a zero divisor in R. In particular, the center of a prime ring is an integral domain. In consequence the center of a primitive ring is an integral domain.*

We claim that, given an integral domain $I \neq (0)$ there is a primitive ring having precisely I as center. For let F be the field of quotients of I and let R be the set of countable matrices over F of the form

$$\begin{pmatrix} A_n & & 0 & \\ & \delta & & \\ & & \cdot & \\ & & & \cdot \\ 0 & & & \delta \\ & & & & \cdot \\ & & & & & \cdot \end{pmatrix}$$

where $\delta \in I$ and A_n is an arbitrary $n \times n$ matrix over F and n is allowed to be any integer. A simple computation shows that the center of R is

$$\left\{ \left\| \begin{pmatrix} \delta & & & & \\ & \cdot & & 0 & \\ & & \cdot & & \\ 0 & & & \cdot & \\ & & & & \delta \\ & & & & & \cdot \\ & & & & & & \cdot \end{pmatrix} \right\| \; \delta \in I \right\}$$

Moreover it is easy to verify that R is primitive. We leave to the reader to construct a primitive ring whose center is (0).

Let R be any ring and let $E(R)$ be the endomorphism ring of the additive group of R. If $a \in R$ we define, as before, $T_a : R \to R$ by $xT_a = xa$ and $L_a : R \to R$ by $xL_a = ax$. For any a, $b \in R$ the maps T_a, L_b are in $E(R)$. Let $B(R)$ be the subring of $E(R)$ generated by all the T_a and L_b for a, $b \in R$. $B(R)$ is often called the *multiplication ring* of R.

Clearly R is a module over $B(R)$; the $B(R)$-submodules of R are merely the two-sided ideals of R. Thus R is irreducible as a $B(R)$-module if and only if R is simple. We ask about the commuting ring of $B(R)$ on R. This is a fundamental invariant of a ring.

DEFINITION. *The centroid of R is the set of elements in $E(R)$ which commute elementwise with $B(R)$.*

LEMMA 2.1.4. *If $R^2 = R$ then the centroid of R is commutative.*

Proof. Suppose that σ, τ are in the centroid of R. For any

$$x, y \in R \ (xy)\sigma = (xT_y)\sigma = (x\sigma)T_y = (x\sigma)y$$

and

$$(xy)(\sigma\tau) = (x\sigma)y)\tau = (yL_{x\sigma})\tau$$
$$= (y\tau)L_{x\sigma} = (x\sigma)(y\tau)$$
$$= (x(y\tau))\sigma = (xy)(\tau\sigma).$$

Therefore $(xy)(\sigma\tau - \tau\sigma) = 0$. Since $R^2 = R$, given $u \in R$, $u = \sum x_i y_i$ hence $u(\sigma\tau - \tau\sigma) = \sum x_i y_i(\sigma\tau - \tau\sigma) = 0$. This gives $\sigma\tau - \tau\sigma = 0$, the required result.

In case R is simple we can say a great deal more about its centroid. To begin with, $R^2 = R$ so the centroid of R is commutative. Moreover, simplicity being equivalent with the irreducibility of R as a $B(R)$-module, by Schur's lemma the centroid must be a division ring. Combining these two we have that the centroid of a simple ring is a field. R is in a natural way an algebra over this field, hence every simple ring is a simple algebra over its centroid. Suppose, in addition, that the center Z of R is not (0). If $0 \neq z \in Z$ then Rz is a nonzero two-sided ideal of R so must be R. This quickly yields that R has a unit element 1. Let $\mathcal{Z}$ denote the centroid of R; it is immediate from definition that $Z \subset \mathcal{Z}$. Let $\sigma \in \mathcal{Z}$; for any $r \in R$, $(1r)\sigma = (1T_r)\sigma = (1\sigma)T_r = (1\sigma)r$. If $a = 1\sigma$ we have shown $r\sigma = ar$. Similarly $r\sigma = (r1)\sigma = (1L_r)\sigma = (1\sigma)L_r = ra$ hence $ra = ar$ for all $r \in R$. Therefore $a \in Z$. Since $r\sigma = ar = ra$ we have that $r(\sigma - a) = 0$. However $\mathcal{Z}$ is a field, $\sigma - a \in \mathcal{Z}$ and R is an algebra over $\mathcal{Z}$, these together tell us that $\sigma - a = 0$ and so $\sigma \in Z$. We have shown that $Z = \mathcal{Z}$ hence, in particular, Z must be a field. We summarize all these remarks in

THEOREM 2.1.5. *If R is a simple ring then its centroid is a field and R is an algebra over this field. If, in addition, the center of R is not (0) then the center and the centroid of R coincide.*

In the last part of this section we concentrate on a

very famous theorem of Wedderburn. It is the corner stone of many things done in algebra. From it comes out the whole theory of group representation. In fact there are very few places in algebra—at least where noncommutative rings are used—where it fails to make its presence felt. Effectively it is a corollary to Theorem 2.1.2. Wedderburn proved the result for finite-dimensional simple algebras, Artin extended it to simple Artinian rings.

THEOREM 2.1.6 (Wedderburn-Artin). *Let R be a simple Artinian ring. Then R is isomorphic to D_n, the ring of all $n \times n$ matrices over the division ring D. Moreover n is unique, as is D up to isomorphism. Conversely, for any division ring D, D_n is a simple Artinian ring.*

Proof. We first show that R is primitive. Since R is Artinian, $J(R)$ is nilpotent; because $R^2 = R$, R is not nilpotent. Hence $J(R) \neq R$, so must be (0). As a semisimple simple ring, R must be primitive. Let M be a faithful irreducible module for R. M is a vector space over the division ring $D = C(M)$ the commuting ring of R on M.

We claim that M is finite-dimensional over D. For if $v_1, \cdots, v_n, \cdots$ in M are linearly independent over D let $\rho_m = \{x \in R \mid v_i x = 0 \text{ for } 0 = 1, 2, \cdots, m\}$. Clearly $\rho_1 \supset \rho_2 \supset \cdots \supset \rho_m \cdots$ is a descending chain of right ideals of R. By Theorem 2.1.2 this is a *properly* descending chain. Since R is Artinian it has no properly descending infinite chain of right ideals. The net result of this is that $\rho_n = (0)$ for some n; but then v_{n+1} is linearly dependent on $v_1, \cdots, v_n$ and thus proves that M is finite-dimensional over D.

As we remarked earlier, in a finite dimensional context density means the set of *all* linear transformations. Hence R being dense on the n-dimensional space M over D we conclude that R is isomorphic to D_n.

Now to the uniqueness of n and D. This amounts to proving that if $D_m \approx \Delta_n$, D, Δ division rings then $m = n$ and $D \approx \Delta$ (recall $\approx$ indicates "isomorphic to").

Let

$$e = \begin{pmatrix} 1 & 0 & \cdots & 0 \\ 0 & 0 & \cdots & 0 \\ \cdot & \cdots & \cdots & \cdots \\ 0 & 0 & \cdots & 0 \end{pmatrix} \in D_m$$

and let $f = \phi(e)$ where ϕ is the isomorphism of D_m onto Δ_n. Since eD_m is a minimal right ideal of D_m, $f\Delta_n$ is a minimal right ideal of Δ_n. By a change of basis, so an automorphism of Δ_n, f can be brought to the form

$$\begin{pmatrix} I_r & 0 \\ 0 & 0 \end{pmatrix}$$

where I_r is the $r \times r$ unit matrix. Since $f\Delta_n$ is a minimal right ideal of Δ_n this forces $r = 1$ (prove!). Thus, without loss of generality,

$$f = \begin{pmatrix} 1 & 0 & \cdots & 0 \\ 0 & 0 & \cdots & 0 \\ \cdot & \cdot & & \cdot \\ 0 & 0 & \cdots & 0 \end{pmatrix}.$$

Now $D \approx eD_m e \approx f\Delta_n f \approx \Delta$ proving the isomorphism of D and Δ. Also, eD_m is m-dimensional over D, $f\Delta_n$ is n-dimensional over Δ; being isomorphic we get $m = n$.

We leave the proof of the converse to the theorem to the reader.

Wedderburn's theorem has implications of great import in many specific cases of Artinian rings. To begin with, Theorem 1.4.4 asserted that any semisimple Artinian ring is a direct sum of a finite number of simple

Artinian rings. Combined with Theorem 2.1.6 this gives us a definitive structure theorem for these rings.

THEOREM 2.1.7. *If R is a semisimple Artinian ring then $R \approx \Delta_{n_1}^{(1)} \oplus \cdots \oplus \Delta_{n_k}^{(k)}$ where the $\Delta^{(i)}$ are division rings and where $\Delta_{n_i}^{(i)}$ is the ring of all $n_i \times n_i$ matrices over $\Delta^{(i)}$.*

Are there circumstances in which we can say more, in which we can identify the Δ's more precisely? One such is the case of a finite-dimensional simple algebra over an algebraically closed field. We need a preliminary definition and lemma.

DEFINITION. *Let A be an algebra over a field F; $a \in A$ is said to be algebraic over F if there is a nonzero polynomial $p(x) \in F[x]$ such that $p(a) = 0$. A is said to be an algebraic algebra over F if every $a \in A$ is algebriac over F.*

Note that if A is finite-dimensional over F then it is algebraic over F; for if $a \in A$ and $n = \dim_F A$ the elements $a, a^2, \cdots, a^{n+1}$, being $n+1$ in number, must be linearly dependent over F. Thus $\alpha_1 a + \cdots + \alpha_{n+1} a^{n+1} = 0$ where the $\alpha_i \in F$ are not all 0. In view of this a satisfies the nonzero polynomial $p(x) = \alpha_1 x + \cdots + \alpha_{n+1} x^{n+1}$ in $F[x]$.

LEMMA 2.1.5. *Let F be an algebraically closed field. If D is a division algebra algebraic over F then $D = F$.*

Proof. Let $a \in D$; then $p(a) = 0$ for some $p(x) \in F[x]$. Since D is a division algebra over F, F is contained in the center of D. Since F is algebraically closed, $p(x) = \Pi(x - \lambda_i), \lambda_i \in F$ whence $0 = p(a) = \Pi(a - \lambda_i)$. Being in a division ring we conclude that $a - \lambda_i = 0$ for some i, hence $a = \lambda_i$ is in F. Thus $D = F$.

With this lemma at our disposal Theorems 2.1.6 and 2.1.7 assume a very nice form for semisimple finite-dimensional algebras over algebraically closed fields.

THEOREM 2.1.8. *Let F be an algebraically closed field and let A be a finite-dimensional semisimple algebra over F. Then $A \approx F_{n_1} \oplus \cdots \oplus F_{n_k}$.*

Proof. It is enough to show this for a simple algebra A. But then $A \approx \Delta_n$ where Δ is a division algebra finite-dimensional over F. By the lemma $\Delta = F$ and so $A \approx F_n$.

One should note an easy consequence of Theorem 2.1.8. Clearly the center of a direct sum is the direct sum of the centers. Also the center of F_{n_i} is one-dimensional over F (being FI_{n_i}, I_{n_i} the $n_i \times n_i$ unit matrix). Thus $k = \dim_F Z$. In other words,

COROLLARY. *If A is as in Theorem 2.1.8 then the number of simple direct summands for A equals the dimension of the center of A over F.*

This remark will stand us in good stead when we study the representations of finite groups.

Another immediate corollary of Theorem 2.1.8 is the structure of the group algebra of a finite group. Since it is of great independent interest and importance we single it out as

THEOREM 2.1.9. *Let G be a finite group of order $o(G)$ and let F be an algebraically closed field of characteristic 0 or $p \nmid o(G)$. Then $F(G) \approx F_{n_1} \oplus \cdots \oplus F_{n_k}$.*

Proof. As we saw in Maschke's Theorem (Theorem 1.4.1), under the hypothesis given, $F(G)$ is semisimple. The result then follows from Theorem 2.1.8.

In particular, when F is the field of complex numbers $F(G)$ has this highly desirable form. We shall exploit it heavily to derive much information about the representations and characters of finite groups.

2. Semisimple rings Having given a fairly satisfactory description of primitive rings in the previous section, we now try to tie the structure of semisimple rings to that of primitive ones. In order to do so we first generalize the concept of direct sum.

Let us recall that by the *direct product* (or *complete direct sum*) of the rings R_γ, γ in some index set I, we mean the set $\Pi_{\gamma \in I} R_\gamma = \{f: \ I \to \cup_{\gamma \in I} R_\gamma | f(\gamma) \in R_\gamma$ all $\gamma \in I\}$. We give a ring structure to

$$\prod_{\gamma \in I} R_\gamma \text{ by defining } (f+g)(\gamma)$$

$$= f(\gamma) + g(\gamma) \text{ and } (fg)(\gamma) = f(\gamma)g(\gamma).$$

Let π_γ be the projection of $\Pi_{\gamma \in I} R_\gamma$ onto R_γ.

DEFINITION. *R is said to be a subdirect sum of the rings* $\{R_\gamma\}_{\gamma \in a}$ *if there is a monomorphism* $\phi: R \to \Pi_{\gamma \in I} R_\gamma$ *such that* $R\phi\pi_\gamma = R_\gamma$ *for each* $\gamma \in I$.

The following result is straightforward; we leave its proof to the reader.

LEMMA 2.2.1. *Let R be a ring and let* $\phi_\gamma: R \to R_\gamma$ *be homomorphisms of R onto rings* R_γ. *Let* $U_\gamma = \text{Ker } \phi_\gamma$. *Then R is a subdirect sum of the* R_γ *if and only if* $\cap_\gamma U_\gamma = 0$.

DEFINITION. *R is said to be subdirectly irreducible if the intersection of all its nonzero ideals is not zero.*

This merely says that R has no nontrivial representation as a subdirect sum.

LEMMA 2.2.2. *Every ring can be represented as a subdirect sum of subdirectly irreducible rings.*

Proof. For $a \neq 0$ in R let U_a be an ideal of R maximal with respect to the exclusion of a. By Zorn's Lemma such a U_a exists. Clearly $\bigcap_{0 \neq a \in R} U_a = (0)$ so that R is a subdirect sum of the rings R/U_a. Now $a + U_a \neq 0$ is in every nonzero ideal of R/U_a, hence R/U_a is subdirectly irreducible.

We give one more representation theorem in terms of general subdirect sums.

LEMMA 2.2.3. *Let R be a ring having no nonzero nil ideals. Then R is a subdirect sum of prime rings.*

Proof. Let $a \in R$ be non-nilpotent and let W_a be an ideal of R maximal with respect to the exclusion of all a^n. If A, B are ideals of R such that $AB \subset W_a$ and $A \not\subset W_a$, $B \not\subset W_a$ then $a^{n_1} \in W_a + A$, $a^{n_2} \in W_a + B$ hence

$$a^{n_1 + n_2} \in (W_a + A)(W_a + B) \subset W_a + AB \subset W_a,$$

a contradiction. Thus W_a is a prime ideal of R and $R_a = R/W_a$ is a prime ring. As a runs over the non-nilpotent elements of R, $\bigcap W_a$ is a nil ideal, hence is (0). R is thus a subdirect sum of the R_a which are prime.

Note that R_a actually has a further property. If $\bar{a}$ denotes the image of a in R_a and if $\overline{U} \neq (0)$ is an ideal of R_a then $\bar{a}^{n(\overline{U})} \in \overline{U}$, that is, the powers of $\bar{a}$ fall in all non-zero ideals of R_a.

Now let R be a semisimple ring. As was shown in Theorem 1.2.1, for any ring B, $J(B) = \bigcap(\rho : B)$ where ρ ranges over the maximal regular right ideals of B. Since R is semisimple $\bigcap(\rho : R) = (0)$. By Lemma 2.2.1 R is a subdirect sum of the $R/(\rho : R)$; by Theorem 2.1.1, for instance, $R/(\rho : R)$ is primitive. Therefore R is a subdirect sum of primitive rings.

On the other hand suppose that R is a subdirect sum of the rings $R_\phi = R/U_\phi$. Therefore $\bigcap U_\phi = (0)$. If the rings R_ϕ are all primitive then they are semisimple. Since $J(R)$ maps into a quasiregular ideal of R_ϕ it must map into (0). Thus $J(R) \subset U_\phi$ for each ϕ, hence $J(R) \subset \cap U_\phi$ $= (0)$ proving that R is semisimple. We have proved

THEOREM 2.2.1. *R is semisimple if and only if it is isomorphic to a subdirect sum of primitive rings.*

Since a commutative primitive ring is a field we have the following

COROLLARY. *A commutative semisimple ring is a subdirect sum of fields.*

The structure theorem for semisimple rings just given is a very loose one. To begin with knowing that a ring is primitive does not tie it down very much—there can be very weird dense rings of linear transformations on a vector space. Hence the pieces in the subdirect sum decomposition can be wild and strange. Added to this is the looseness implicit in the concept of subdirect sum. For instance, uniqueness is not one of its attributes. The ring of integers can be realized as the subdirect sum of J_p, the integers modulo the prime p, over any infinite set of primes.

However, it would be completely unrealistic to expect a very tight structure theorem. The blanket of semi-simplicity covers many strange and disparate bedmates. Yet despite all these shortcomings these structure theorems have been used to great effect to prove results about general rings.

A program for attacking a given problem in ring theory is the following: first prove the theorem for division rings—this may involve you in arithmetic questions in field theory—then pass to primitive rings and via Theorem 2.1.4 induct the result into matrix rings. Tie these together to get the result for semisimple rings. Now we know the theorem for $R/J(R)$; all that remains is to climb back through the radical to get the result for R. These steps may all be beset with difficulties, especially the first and last ones. It must be so, for in general the technique fails. Yet it is a specific program of attack and fortunately it has worked often enough to have justified its existence.

Let's see how this all works in a specific instance. The problem we study now is special, contrived and of little interest. Besides, it is a very special case of a general theorem we prove later. But all this notwithstanding it does illuminate well the general procedure. Later we shall see the same type of mechanism applied to interesting and intrinsic situations.

Let R be a ring in which, for any $a, b \in R$, $(ab - ba)^3 = (ab - ba)$ (a condition admittedly rather special). We want to prove that R is commutative.

First let R be a division ring; if $ab - ba \neq 0$ then $(ab - ba)^3 = ab - ba$ yields $(ab - ba)^2 = 1$ and so $ab - ba = \pm 1$. At any rate, for all $a, b \in R$ we get $ab - ba \in Z$, the center of R. If $ab - ba = \alpha \neq 0$ then since $a(ab) - (ab)a \in Z$ we get $\alpha a \in Z$.

From $\alpha \neq 0$ we end up with $a \in Z$, giving $ab - ba = 0 \neq \alpha$. This contradiction assures us that when R is a division ring then it is in fact a commutative field.

Now let R be primitive; if R is not a division ring, using Theorem 2.1.4, D_n for some $n > 1$ is a homomorphic image of a subring of R where D itself is a division ring. Since $(ab - ba)^3 = ab - ba$ holds in subrings of R and is preserved under homomorphism we have that this identity holds in D_n, $n > 1$. But this is patently false as we see using

$$a = \begin{pmatrix} 1 & 0 & \cdots & 0 \\ 0 & 0 & \cdots & 0 \\ \cdot & \cdot & & 0 \\ 0 & 0 & \cdots & 0 \end{pmatrix}, \quad b = \begin{pmatrix} 0 & 1 & 0 & \cdots & 0 \\ 0 & 0 & 0 & \cdots & 0 \\ \cdot & \cdot & \cdot & & 0 \\ 0 & 0 & 0 & \cdots & 0 \end{pmatrix}$$

for here $ab - ba = b$ and since $b^2 = 0$, $0 = (ab - ba)^3$

$\neq ab - ba = b$. We have established that if R is primitive it must be a division ring, hence must be commutative.

Any semisimple ring R is a subdirect sum of primitive rings R_ϕ each of which is a homomorphic image of R. The relation $(ab - ba)^3 = ab - ba$ persists in R_ϕ forcing it to be commutative. Hence R, as a subring of a direct sum of commutative rings, must be commutative.

Now let R be a general ring satisfying our identity. Therefore $R/J(R)$, as a semisimple ring enjoying the hypothesis, must be commutative. But then for all a, $b \in R$ the commutator $ab - ba$ must be in $J(R)$. Since $(ab - ba)^3 = ab - ba$ and $ab - ba \in J(R)$, we are forced to conclude that $ab - ba = 0$. (The reason for this is that if $ux = u$ with $x \in J(R)$ then $u = 0$ (prove!).) In this way what we set out to prove has been proved.

3. Applications of Wedderburn's Theorem.

As we indicated earlier, the theorem of Wedderburn which so explicitly gives the structure of simple and semisimple Artinian rings has numerous applications in a variety of places in algebra. One large such application is to the study of the representations in matrices of finite groups; we shall presently go into this in some detail. We now give several applications in other directions. One of these ties in the global nature of certain algebras with local properties of the elements in a basis. Another proves the Burnside problem for groups of matrices.

We begin with a delightful theorem also due to Wedderburn—it occurs in one of the last papers he published.

THEOREM 2.3.1. *Let A be a finite dimensional algebra over a field F. Suppose that A has a basis over F consisting of nilpotent elements. Then A itself must be nilpotent.*

Proof. Let $\{u_1, \cdots, u_n\}$ be a basis of A over F such that each u_i is nilpotent. We assert first that without loss of generality F may be assumed to be algebraically closed. For let $\overline{F}$ be the algebraic closure of F and let $\overline{A} = A \otimes_F \overline{F}$. Now $\overline{A}$ has a nilpotent basis $\{u_i \otimes 1\}$ over $\overline{F}$ so if we could prove that $\overline{A}$ were nilpotent we would have $A \otimes 1$, which is isomorphic to A, nilpotent.

We thus proceed assuming that F is algebraically closed. Our proof is by induction on dim A.

If dim $A = 1$ then $A = Fu_1$ and since u_1 is nilpotent we immediately have that A is nilpotent.

Suppose that the theorem is true for all algebras B such that dim $B < n$. Let dim $A = n$ where A has a basis consisting of nilpotent elements. Let $J(A)$ be the radical of A; if $J(A) = A$ then A is nilpotent and we would be done. If $J(A) \neq A$ we want to reach a contradiction. Now if $J(A) \neq (0)$ then dim $A/J(A) < n$ and $A/J(A)$ has a nilpotent basis $\{u_i + J(A)\}$ so by induction is nilpotent. This is nonsense since $A/J(A)$ is semisimple. We must therefore assume that $J(A) = (0)$, that is, that A is semisimple. By Theorem 1.4.4, $A = A_1 \oplus \cdots \oplus A_k$ when the A_i are simple algebras. If $k > 1$ then A_1 is of smaller dimension than A and is a homomorphic image of A so has a nilpotent basis; by induction A_1 is nilpotent which violates the semisimplicity of A. Hence $k = 1$, which is to say that A is simple.

Since F is algebraically closed and A is a finite dimensional simple algebra over F, by Theorem 2.1.8, $A \approx F_m$ for some m. Thus F_m has a basis $a_1, \cdots, a_{m^2}$ of nilpotent elements. Each a_i as a nilpotent matrix has trace 0 so any linear combination of these has trace 0. The elements $a_1, \cdots, a_{m^2}$ form a basis of F_m hence

$$\begin{pmatrix} 1 & 0 & \cdots & 0 \\ 0 & 0 & \cdots & 0 \\ \cdot & \cdot & & \cdot \\ 0 & 0 & \cdots & 0 \end{pmatrix}$$

as a linear combination of $a_1, \cdots, a_{m^2}$ has trace 0. Since its trace is 1 this is manifestly false. The theorem has now been proved.

Instead of considering algebras over fields one could consider algebras over arbitrary commutative rings. For this case Herstein, Procesi and Small have shown: Let A be an algebra over a commutative ring C which is finitely generated as a C-module; if A has a set of generators each of which is nilpotent then A is nilpotent. This pushes the theorem of Wedderburn, that we have just proved, to a limit in one direction. We could try others. For instance we could ask: let R be an Artinian ring in which every element is a sum of nilpotent ones; is R then nilpotent? The question was posed by Kaplansky and shown to be in the negative by Harris.

We apply this result to group algebras. Let G be a group of order p^m, p a prime and let F be a field of characteristic p. We saw earlier that $F(G)$ is not semisimple, but what exactly is its radical? This is

THEOREM 2.3.2. *If G has order p^m and F is a field of characteristic p then the radical of $F(G)$ has dimension $p^m - 1$, having as a basis the elements $g - 1$ where $g \in G$. More succinctly, $J(F(G)) = \{\Sigma \alpha_i g_i \in F(G) \,|\, \Sigma \alpha_i = 0\}$.*

Proof. Let $U = \{x = \Sigma \alpha_i g_i \in F(G) \,|\, \Sigma \alpha_i = 0\}$. U is an ideal of $F(G)$—it is called the *augmentation ideal*—which is of dimension $p^m - 1$ over F and has the $g - 1$, $g \in G$, as basis. Now for every $g \in G$, $(g-1)^{p^m} = g^{p^m} - 1 = 1 - 1 = 0$; since U has a basis of nilpotent elements U must be nilpotent. As a nilpotent ideal $U \subset J(F(G))$. However, since U is a maximal ideal of $F(G)$ we infer that $U = J(F(G))$. This is the theorem.

The Wedderburn theorem is also a useful tool for the study of groups of matrices. Recall that a set S of linear transformations on a vector space V is said to be irreducible if no nontrivial subspace of V is invariant with respect to S. Otherwise S is said to be reducible. If S is a set of $n \times n$ matrices over a field then S is reducible if there exists an invertible matrix P such that

$$PaP^{-1} = \begin{pmatrix} a_{11} & a_{12} \\ 0 & a_{22} \end{pmatrix}$$

for all $a \in S$; here a_{11}, a_{22} are square submatrices. By a semigroup of matrices we shall mean a nonempty set of matrices closed under matrix multiplication.

THEOREM 2.3.3. *Let S be an irreducible semigroup of $n \times n$ matrices over a field F. Suppose that tr a, the trace of a, takes on k distinct values as a ranges over S. Then S has at most k^{n^2} elements.*

Proof. Let $\overline{F}$ be the algebraic closure of F; then clearly $S \subset F_n \subset \overline{F}_n$. In other words, we may assume that F is algebraically closed.

Let $A = \left\{ \sum \alpha_i a_i \mid \alpha_i \in F, \ a_i \in S \right\}$ be the linear span of S. Since S is a semigroup, A is a subalgebra of F_n. Moreover A acts irreducibly and faithfully on V, the set of n-tuples over F, since S itself is already irreducible. The commuting ring of A on V being a finite dimensional division algebra over F must be F itself. By Wedderburn's theorem $A = F_n$ follows.

Since S spans $A = F_n$ over F there must be matrices $a_1, \cdots, a_{n^2}$ in S which form a basis of F_n over F. Let $\sigma_1, \cdots, \sigma_k$ be the k values assumed by the traces of the elements of S. If $x \in S$ then tr $a_1 x, \cdots,$ tr $a_{n^2} x$ is an n^2-tuple of elements each of which is one of $\sigma_1, \cdots, \sigma_k$. Since there are k^{n^2} such n^2-tuples we will be done if we can show that the system of equations tr $a_1 x = \beta_1, \cdots,$ tr $a_{n^2} x = \beta_{n^2}$, $\beta_i \in F$ has at most one solution in F_n. However, these equations are linear, so it is enough to show that tr $a_i x = 0$, $i = 1, \cdots, n^2$ has only $x = 0$ as solution.

Now the a_i are a basis of F_n so if tr $a_i x = 0$ for all i then tr $yx = 0$ for all matrices y in F_n. Since the trace is a non-degenerate bilinear form this indeed forces $x = 0$.

This result enables us to limit the nature of matrix groups of a certain kind over fields of characteristic 0.

THEOREM 2.3.4. *Let G be a multiplicative group of $n \times n$ matrices over a field F of characteristic 0. Suppose that there is a positive integer N such that $a^N = 1$ for every $a \in G$. Then G is a finite group.*

Proof. Since $F \subset \overline{F}$, the algebraic closure of F, and since $G \subset F_n \subset \overline{F}_n$, we may assume without loss of generality that F is algebraically closed.

We proceed by induction on n. If $n = 1$ then G is a multiplicative subgroup of the nonzero elements of F; since in F the equation $x^N = 1$ has at most N roots, G must be finite.

We suppose the result to be true for multiplicative groups in F_m for $m < n$ and that $G \subset F_n$ satisfies the hypothesis of the theorem.

If $a \in G$ then from $a^N = 1$ all the characteristic roots of a are Nth roots of unity, hence are finite in number. Since tr a is the sum of its characteristic roots we can only get a finite number k of distinct traces in G. Thus if G is an irreducible group of matrices by the preceding theorem G has at most k^{n^2} elements so is finite.

Suppose then that G is reducible. By an appropriate change of basis we can assume that every $a \in G$ is of the form

$$\begin{pmatrix} a_1 & b_1 \\ 0 & a_2 \end{pmatrix}$$

where $a_1 \in F_m$, $a_2 \in F_{n-m}$ and $0 < m < n$. A simple check verifies that

$$G_1 = \left\{ a_1 \in F_m \,\middle|\, \exists a \in G, \; a = \begin{pmatrix} a_1 & b_1 \\ 0 & a_2 \end{pmatrix} \right\}$$

and

$$G_2 = \left\{ a_2 \in F_{n-m} \,\middle|\, \exists a \in G, \; a = \begin{pmatrix} a_1 & b_1 \\ 0 & a_2 \end{pmatrix} \right\}$$

are groups of matrices in F_m and F_{n-m} respectively. Moreover one easily verifies that $a_i \in G_i (i = 1, 2)$ satisfies $a_i^N = 1$. By our induction G_1 and G_2 are finite groups. Given $a_1 \in G_1$ and $a_2 \in G_2$ we claim that *there is at most one $m \times (n - m)$ matrix b_1* such that

$$\begin{pmatrix} a_1 & b_1 \\ 0 & a_2 \end{pmatrix}$$

is in G. For suppose that

$$\begin{pmatrix} a_1 & b_1 \\ 0 & a_2 \end{pmatrix} \quad \text{and} \quad \begin{pmatrix} a_1 & c_1 \\ 0 & a_2 \end{pmatrix}$$

are both in G with $c_1 \neq b_1$. Then

$$\begin{pmatrix} a_1 & c_1 \\ 0 & a_2 \end{pmatrix}^{-1} = \begin{pmatrix} a_1^{-1} & -a_1^{-1} c_1 a_2^{-1} \\ 0 & a_2^{-1} \end{pmatrix}$$

is in G hence

$$\begin{pmatrix} a_1 & b_1 \\ 0 & a_2 \end{pmatrix} \begin{pmatrix} a_1^{-1} & -a_1^{-1} c_1 a_2^{-1} \\ 0 & a_2^{-1} \end{pmatrix}$$

is in G. This computes out to be:

$$\begin{pmatrix} I_m & (b_1 - c_1) a_2^{-1} \\ 0 & I_{n-m} \end{pmatrix} \in G.$$

As an element of G it must be of finite period. *Now in characteristic* 0

$$\begin{pmatrix} I_m & w \\ 0 & I_{n-m} \end{pmatrix}$$

has finite period only if $w = 0$. Thus $(b_1 - c_1)a_2^{-1} = 0$, and so $b_1 = c_1$.

Therefore, having shown that for $a_1 \in G_1$, $a_2 \in G_2$ there is at most one element b_1 with

$$\begin{pmatrix} a_1 & b_1 \\ 0 & a_2 \end{pmatrix}$$

in G we know that $o(G) \leqq o(G_1)o(G_2)$. Thus G is a finite group.

Note that the theorem just proved is false in characteristic $p \neq 0$. For let F an infinite field of characteristic p and let

$$G = \left\{ \begin{pmatrix} 1 & \alpha \\ 0 & 1 \end{pmatrix} \middle| a \in F \right\};$$

G is a group, is infinite and yet

$$\begin{pmatrix} 1 & \alpha \\ 0 & 1 \end{pmatrix}^p = \begin{pmatrix} 1 & 0 \\ 0 & 1 \end{pmatrix}.$$

We head for the Burnside problem for matrix groups; before doing so we give several levels of the problem for the general case.

DEFINITION. *A group G is said to be a torsion group if every element in G is of finite order.*

DEFINITION. *A group G is said to be locally finite if every finitely generated subgroup of G is finite.*

Clearly a locally finite group is a torsion group. What about the other way? This is precisely the Burnside problem. We state two possible Burnside problems.

1. *General Burnside Problem:* Is every torsion group locally finite?

2. *Bounded Burnside Problem:* Let G be a torsion group in which $x^N = 1$ for all $x \in G$ N a fixed positive integer. Is G then locally finite?

The situation for these problems stands as follows.

1. As a result of the work of Golod and Shafarevitch (which we do at the end of the book) the general Burnside problem is answered in the negative.
2. Novikov announced the existence of an infinite group G_N generated by two elements in which $x^N = 1$ holds for all $x \in G$. This for any $N \geqq 72$. The proof has, as yet, never been published in detail.

However, for matrix groups Burnside himself settled the general Burnside problem in the affirmative. Note that in Theorem 2.3.4 we showed that in characteristic 0 a torsion group of matrices in which the orders of the elements are bounded is finite—this even without the assumption that the group is finitely generated. We now go about settling these questions for matrix groups. The proof we follow is along the lines of that given by Kaplansky in his "Notes on Ring Theory."

LEMMA 2.3.1. *Suppose that G is a group, N a normal subgroup of G such that both N and G/N are locally finite. Then G is locally finite.*

Proof. Let $g_1, \cdots, g_n$ be a finite set of elements of G; we wish to show that they generate a finite subgroup of G. If $\bar{g}_1, \cdots, \bar{g}_n$ denote their images in G/N then, by assumption, these generate a finite subgroup of G/N. Let this subgroup be $\{\bar{g}_1, \cdots, \bar{g}_n, \cdots, \bar{g}_t\}$ and let $g_{n+1}, \cdots, g_t$ be inverse images of $\bar{g}_{n+1}, \cdots, \bar{g}_t$ respectively in G. For any i, j, $g_i g_j = u_{ij} g_k$ for some k and some elements u_{ij} in N. Let U be the subgroup of N generate by all the u_{ij}; the local finiteness of N implies that U is a finite group. Given any three g_i, g_j, g_m then $g_i g_j g_m$

$= u_{ij}g_kg_m = u_{ij}u_{km}g_w$, so is of the form ug_w with $u \in U$. Similarly any word in the g_i's is of the form $u\, g_w$ with $u \in U$, $1 \leq w \leq t$. Hence the $g_1, \cdots, g_t$ generate a group of order at most $to(U)$, that is, a finite group. This establishes the lemma.

This lemma enables us to settle quite easily the Burnside problem for solvable groups.

LEMMA 2.3.2. *A solvable torsion group is locally finite.*

Proof. Let G be a solvable torsion group. By the solvability of G we can find subgroups G_i where G_i is normal in G_{i-1} and G_{i-1}/G_i is abelian and where $G = G_0 \supset G_1 \supset G_2 \supset \cdots \supset G_i = (1)$. An abelian torsion group is clearly locally finite; applying Lemma 2.3.1 we see that we can climb up this chain to get that G is locally finite.

We need one more collateral result before returning to the Burnside problem.

LEMMA 2.3.3. *A group of triangular matrices over a field is solvable.*

Proof. Because a subgroup of a solvable group is solvable it is enough to show that the group of invertible triangular matrices is solvable. To see this let:

$$G = G_0 = \left\{ \begin{pmatrix} \alpha_1 & & * \\ & \ddots & \\ 0 & & \alpha_n \end{pmatrix} \middle| \alpha_i \neq 0 \right\}, \qquad G_1 \left\{ \begin{pmatrix} 1 & & * \\ & \ddots & \\ 0 & & 1 \end{pmatrix} \right\},$$

$$G_2 = \left\{ \begin{pmatrix} 1 & 0 & & * \\ & \ddots & \ddots & \\ & & \ddots & 0 \\ 0 & & & 1 \end{pmatrix} \right\}, \qquad G_3 = \left\{ \begin{pmatrix} 1 & 0 & 0 & & * \\ & \ddots & \ddots & \ddots & \\ & & \ddots & \ddots & 0 \\ & & & \ddots & 0 \\ 0 & & & & 1 \end{pmatrix} \right\},$$

and so on. Each G_i is normal in G_{i-1}, G_{i-1}/G_i is abelian and $G_n = (1)$. Thus the group of triangular matrices is solvable.

We now come to the crucial step in the disposition of the Burnside problem for matrix groups. Here we tie in the finite generation with the orders of the elements in the torsion group. This is

LEMMA 2.3.4. *Let G be a finitely generated torsion group of matrices over a field F. Then there exists a positive integer N such that $\alpha^N = 1$ for any α which is a characteristic root of any element of G.*

Proof. Let $G \subset F_k$ be generated by $a_1, \cdots, a_r$. If P is the prime field of F let F_1 be the field obtained by adjoining all the entries of $a_1, \cdots, a_r$ to P. Clearly every element in G has entries in F_1. Since F_1 is finitely generated over P we can find a subfield K of F_1 which is purely transcendental over P and such that $[F_1 : K] = m$ is finite. Using the regular representation of F_1 over K we can write F_1 as a set of $m \times m$ matrices over K. Substituting these matrices for the entries of F_1 in the elements of G, we realize G as a group of $mk \times mk$ matrices over the field K. In other words we may consider that $G \subset K_t$ for some t where K is a finitely generated purely transcendental extension of P.

Let α be a characteristic root of any element of G; since G is a torsion group α is a root of unity so is algebraic over the prime field P. Since $G \subset K_t$, by the Cayley-Hamilton Theorem any element of G satisfies a polynomial over K of degree t therefore the characteristic roots of the elements of G satisfy polynomials over K of degree t. Since α is such and since K is purely transcendental over P we deduce that α is algebraic over P of degree at most t.

The argument now divides according to the characteristic of P.

1. If P is the field of p elements, p a prime, then $[P(\alpha):P] \leq t$ as we have just seen, hence $P(\alpha)$ is a finite field having p^k elements with $k \leq t$. Thus $\alpha^{k-1} = 1$; since $k \leq t$ for all the characteristic roots α of G the result follows.

2. If the characteristic of P is 0, P is the rational field and all the characteristic roots of the elements of G lie in extension fields of degree at most t over P. Since a primitive mth root of unity has as its minimal polynomial the cyclotomic polynomial which is irreducible and of degree $\phi(m)$, the Euler ϕ-function, and since $\phi(m)$ goes to infinity with m we conclude that there are only a finite number of roots of unity present. The elements of G thus are such that all their characteristic roots come from a finite set of roots of unity. Hence there is a positive integer N such that $\alpha^N = 1$ for all such α.

We have all the required pieces to prove

THEOREM 2.3.5 (Burnside). *A torsion group of matrices over a field is locally finite.*

Proof. Let $G^* \subset F_n$ be a torsion group of matrices. We go by induction on n.

If $n = 1$ then $G^* \subset F$ and so the result is trivial. Suppose the result true for matrices of order less than n.

If $G^* \subset F_n$ is a torsion group of matrices let G be a finitely generated subgroup of G^*. We would like to prove that G is finite.

By Lemma 2.3.4 we have that there exists an integer $N > 0$ such that $\alpha^N = 1$ for any α which is a characteristic root of an element of G. In consequence, tr g as g runs over G, takes on only a finite number of values. If G should be an irreducible group of matrices it would be finite by applying Theorem 2.3.3. So suppose that G is reducible. By a change of basis we can assume that every $g \in G$ is of the form

$$\begin{pmatrix} g_1 & 0 \\ b_1 & g_2 \end{pmatrix}$$

where $g_1 \in F_m$, $g_2 \in F_{n-m}$ for $0 < m < n$. The set G_1 of g_1 arising this way is a torsion group of $m \times m$ matrices over F so by induction it is locally finite. (In fact it is finitely generated so is actually even finite.) Similarly G_2, the set of g_2 arising, is a locally finite group.

Given

$$\begin{pmatrix} g_1 & 0 \\ b_1 & g_2 \end{pmatrix} \text{ in } G \text{ map it onto } \begin{pmatrix} g_1 & 0 \\ 0 & g_2 \end{pmatrix};$$

this map ϕ is clearly a homomorphism of G into a locally finite group. Moreover, Ker ϕ is a subgroup of the group of triangular matrices and, as a subgroup of a torsion group is itself a torsion group. Invoking Lemma 2.3.2 we deduce that Ker ϕ is locally finite. Knowing that both Ker ϕ and $G/\text{Ker } \phi$ are locally finite we have by Lemma 2.3.1 that G itself is locally finite. Since G is finitely generated it must therefore be finite. We have proved that G^* is locally finite. Thus the theorem is established.

The theorem is capable of a rather wide extension. This was done by Herstein and Procesi. Their result reads as follows: Let R be a ring satisfying a polynomial identity and suppose that $G \subset R$ is a torsion group relative to the product in R; then G is locally finite.

References

1. G. Bergman, A ring primitive on the right but not on the left, *Proc. Amer. Math. Soc.*, 15 (1964) 473–475.

2. B. Harris, Commutators in division rings, *Proc. Amer. Math. Soc.*, 9 (1958) 628–630.

3. E. S. Golod and I. R. Shafarevitch, On towers of class fields, *Izv. Akad. Nauk SSR*, 28 (1964) 261–272.

4. I. N. Herstein, *Theory of rings*, University of Chicago, Math Lecture Notes, 1961.

5. I. N. Herstein and Lance Small, Nil rings satisfying certain chain conditions: an addendum, *Canad. J. Math.*, 18 (1966) 300–302.

6. Nathan Jacobson, Structure of rings, *Amer. Math. Soc. Colloq. Publ.*, 37 (1964).

7. ———, Radical and semi-simplicity for arbitrary rings, *Amer. J. Math.*, 67 (1945) 300–320.

8. Irving Kaplansky, *Notes on ring theory*, Univ. of Chicago, Math Lecture Notes, 1965.

9. Claudio Procesi, On the Burnside problem, *Journal of Algebra*, 4 (1966) 421–426.

10. J. H. M. Wedderburn, On hypercomplex numbers, *Proc. London Math. Soc.*, 6 (1908) 77–117.

11. ———, Note on algebras, *Ann. of Math.*, 38 (1937) 854–856.

CHAPTER 3

COMMUTATIVITY THEOREMS

In the preceding two chapters we laid out a general line of attack on ring-theoretic problems. This procedure is especially efficacious in proving that appropriately conditioned rings are commutative, or almost so. The reason for this lies in the trivial fact that a subring of a direct product of commutative rings is itself commutative. In the theorems to be considered we shall see clearly the role assumed by this general structure theory in the disposition of the problems at hand. Most particularly they illustrate effectively the point made earlier that in this kind of approach the difficulties usually present themselves at two stages, namely in the first step of establishing the result for division rings and in the last one of climbing back through the radical. This area of results together with that of rings with polynomial identities (which we consider later) are probably those parts of ring theory in which the structure theory has had its most successful play.

1. Wedderburn's Theorem and some generalizations. In 1905 Wedderburn proved that a finite division ring is a field. Aside from its own intrinsic beauty this result plays an important role in many diverse parts of algebra —in the theory of group representation and in the theory of algebras to cite two examples. It provides the only known proofs of the fact that in a finite projective plane Desargues' theorem implies that of Pappus'. For us it will serve as the starting point for an investigation of certain kinds of conditions that render a ring commutative.

We begin with

LEMMA 3.1.1. *Let D be a division ring of characteristic $p \neq 0$ and let Z be the center of D. Suppose that $a \in D$, $a \notin Z$ is such that $a^{p^n} = a$ for some $n \geq 1$; then there exists an $x \in D$ such that $xax^{-1} = a^i \neq a$ for some integer i.*

Proof. We define the mapping $\delta: D \to D$ by $x\delta = xa - ax$ for every $x \in D$. Since the characteristic of D is $p \neq 0$ a simple computation yields that $x\delta^p = xa^p - a^p x$. Continuing we obtain that $x\delta^{p^k} = xa^{p^k} - a^{p^k} x$ for all $k \geq 0$.

Let P denote the prime field of Z; since a is algebraic over P, $P(a)$ must be a finite field having p^m elements, say. Hence $a^{p^m} = a$ and so $x\delta^{p^m} = xa^{p^m} - a^{p^m} x = xa - ax = x\delta$ which is to say $\delta^{p^m} = \delta$.

If $\lambda \in P(a)$ then $(\lambda x)\delta = (\lambda x)a - a(\lambda x) = \lambda(xa - ax) = \lambda(x\delta)$ since λ commutes with a. If λI denotes the map of D into D taking x into λx we have that λI commutes with δ for $\lambda \in P(a)$. Now the polynomial $t^{p^m} - t$ factors in $P(a)$ as $\prod_{\lambda \in P(a)} (t - \lambda)$; since λI commutes with δ we have that $0 = \delta^{p^m} - \delta = \prod_{\lambda \in P(a)} (\delta - \lambda I)$. Since $a \notin Z$, $\delta \neq 0$; let $\delta(\delta - \lambda_1 I) \cdots (\delta - \lambda_k I)$ be the shortest product with the $\lambda_i \in P(a)$ which is 0. By the argument above such a product exists and since $\delta \neq 0$, $k \geq 1$ follows. Hence for some $r \neq 0$ in D, $r(\delta(\delta - \lambda_1 I) \cdots (\delta - \lambda_{k-1} I) = w \neq 0$ yet $w(\delta - \lambda_k I) = 0$. That is, $wa - aw = \lambda_k w$ with $\lambda_k \neq 0$ in $P(a)$. Since $w \neq 0$ we get $waw^{-1} = \lambda_k + a \neq a$ is in $P(a)$.

Now $P(a)$ is a finite field so, in it, the only elements whose order is that of a must be appropriate powers of a. But $waw^{-1} \in P(a)$ is such an element, hence $waw^{-1} = a^i \neq a$ for some i. This is the assertion of the lemma.

This lemma provides us with the requisite tool not only to prove Wedderburn's Theorem but to pass from it to a beautiful result due to Jacobson.

THEOREM 3.1.1 (Wedderburn). *A finite division ring is a commutative field.*

Proof. Let D be a finite division ring and Z its center;

D is of characteristic p and has $q = p^n$ elements. We go by induction on q, assuming that all division rings having fewer than q elements are commutative.

If a, $b \in D$ are such that $ab \neq ba$ but $b^t a = ab^t$ then $b^t \in Z$ for $N(b^t) = \{x \in D \mid xa = ax\}$ is a subdivision ring of D, contains a and b hence is not commutative, hence by the induction $N(b^t) = D$.

If $u \in D$ let $m(u)$ be the least positive power such that $u^{m(u)} \in Z$. Pick $a \in D$, $a \notin Z$ such that $r = m(u)$ is minimal; clearly r is a prime. By Lemma 3.1.1 there is an $x \in D$ such that $xax^{-1} = a^i \neq a$. Hence $x^k a x^{-k} = a^{i^k}$; in particular for $k = r-1$, since $i^{r-1} \equiv 1(r)$ we have that $x^{r-1} a x^{-(r-1)} = \lambda a$, $\lambda \in Z$. Since $xa \neq ax$ and $x^{r-1} \notin Z$ (for $r-1 < r$) we conclude that $\lambda \neq 1$. Let $b = x^{r-1}$; we have that $bab^{-1} = \lambda a$, hence $\lambda^r a^r = (bab^{-1})^r = ba^r b^{-1} = a^r$ and so $\lambda^r = 1$. This gives $ab^r = b^r a$ hence $b^r \in Z$. Let $a^r = \alpha \in Z$, $b^r = \beta \in Z$.

We claim that if $u + u_1 b + \cdots + u_{r-1} b^{r-1} = 0$ with the $u_i \in Z(a)$ then each $u_i = 0$. For let $u_0 + u_1 b^{m_1} + \cdots + u_k b^{m_k} = 0$ be a shortest such relation, $m_1 < m_2 < \cdots m_k < r$; conjugating with a and using $a^{-1} b^i a = \lambda^i b^i$ we get $u_0 + \lambda^{m_1} u_1 b^{m_1} + \cdots + \lambda^{m_k} u_k b^{m_k} = 0$. Playing these off against each other and using that $\lambda^r = 1$, $\lambda \neq 1$ and r is a prime we get the result. In particular, the polynomials $t^r - \alpha$, $t^r - \beta$ are the minimal polynomials for a and b respectively over Z. Hence $[Z(a) : Z] = [Z(b) : Z] = r$.

Now b induces an automorphism ϕ on $Z(a)$ via $\phi(x) = bxb^{-1}$ of order r; since $[Z(a) : Z] = r$ and ϕ leaves Z fixed, the powers of ϕ give all automorphisms of $Z(a)$ leaving Z fixed. As is well known and easy (prove!) the finiteness of $Z(a)$ then tells us that every element $u \neq 0$ in Z can be written as $u = x \phi(x) \cdots \phi^{r-1}(x)$ with $x \in Z(a)$. In particular, $\beta^{-1} = y \phi(u) \cdots \phi^{r-1}(y)$ for some $y \in Z(a)$. But

$$(1 - yb)(1 + yb + y\phi(y)b^2 + \cdots + y\phi(y) \cdots \phi^{r-2}(y)b^{r-1}).$$

Thus either $1 - yb = 0$ or $1 + yb + y\phi(y)b + \cdots + y\phi(y) \cdots \phi^{r-2}(y)b^{r-1} = 0$. Since $y \in Z(a)$ we saw above that either of these is impossible. This proves the theorem.

An immediate consequence of Wedderburn's theorem is the following.

LEMMA 3.1.2. *If D is a division ring of characteristic $p \neq 0$ and $G \subset D$ is a finite multiplicative subgroup of D then G is abelian (and so, cyclic).*

Proof. Let P be the prime field of D and let $A = \{ \sum \alpha_i g_i \mid \alpha_i \in P, \ g_i \in G \}$. Clearly A is a finite subgroup of D under addition; moreover, since G is a group under multiplication A is a finite subring of D. Therefore A is a finite division ring, hence is commutative. Since $G \subset A$ we get the lemma.

The theorem of Wedderburn and the preceding lemma allow us to pass to a much wider situation in which commutativity is forced. The final conclusion is seen as a consequence of algebraic conditions on the elements rather than the finiteness of the system.

LEMMA 3.1.3. *Let D be a division ring in which for any $a \in D$ there exists an integer $n(a) > 1$ such that $a^{n(a)} = a$. Then D is a commutative field.*

Proof. Since $2 \in D$ and $2^m = 2$ with $m > 1$ we have that D is of characteristic $p \neq 0$ for some prime p. If D is not commutative then there is an $a \in D$ and $a \notin Z$, the center of D. Let P be the prime field of Z; since a is algebraic over P, $P(a)$ is a finite field with p^s elements, say, hence $a^{p^s} = a$. All the conditions of Lemma 3.1.1 are fulfilled for this a therefore there is an element $b \in D$ with $bab^{-1} = a^i \subset a$. This relation, together with the fact that a and b have finite period, implies that a and b generate a finite multiplicative subgroup G of D. By Lemma 3.1.2,

G is abelian; since a, $b \in G$ and $ab \neq ba$ this is impossible. The lemma has been demonstrated.

The way is now clear to prove a beautiful theorem due to Jacobson. It is a remarkably wide generalization of Wedderburn's theorem.

THEOREM 3.1.2 (Jacobson). *Let R be a ring in which for every $a \in R$ there exists an integer $n(a) > 1$, depending on a, such that $a^{n(a)} = a$ then R is commutative.*

Proof. To begin with, R is semisimple for if $a \in J(R)$ then since $a^{n(a)} = a$ we have $a(1 - a^{n(a)-1}) = 0$; this is impossible, for as $a^{n(a)-1} \in J(R)$, $1 - a^{n(a)-1}$ is "formally" invertible, unless $a = 0$. Hence $J(R) = (0)$.

By Theorem 2.2.1 B is a subdirect sum of primitive rings R_α; each R_α as a homomorphic image of R inherits the condition $a^{n(a)} = a$. Furthermore any subring of R_α and homomorphic image thereof also satisfies our hypothesis.

As a primitive ring, invoking Theorem 2.1.4 either $R_\alpha \approx D_n$ or any D_m, D a division ring, is a homomorphic image of a subring of R_α. Thus if R_α is not isomorphic to D we get that some D_k with $k > 1$ enjoys the property $a^{n(a)} = a$ for $a \in D_k$, $n(a) > 1$. This is clearly false for the element

$$a = \begin{pmatrix} 0 & 1 & 0 & \cdots & 0 \\ 0 & 0 & 0 & \cdots & 0 \\ . & . & . & & . \\ 0 & 0 & 0 & \cdots & 0 \end{pmatrix}$$

which satisfies $a^2 = 0$. Hence R_α is a division ring so must be commutative by Lemma 3.1.3. In this way R is seen as a subdirect sum of commutative rings and so must be commutative.

It is possible to make the passage from Lemma 3.1.3

to Theorem 3.1.2 without the use of structure theory; a simple, elementary argument of this type may be found in [7].

The theorem as proved has one drawback; true enough, it implies commutativity but only very few commutative rings exist which satisfy its hypothesis. For this reason we try to extend it—something we shall do in a variety of ways—in such a manner that the conditions imposed will automatically hold for all commutative rings yet will themselves force the commutativity of the ring in question. The first such extension is

THEOREM 3.1.3. *Let R be a ring such that for every x, $y \in R$ there exists an integer $n(x, y) > 1$ such that $(xy - yx)^{n(x,y)} = (xy - yx)$. Then R is commutative.*

In order to prove this theorem we must first establish it in a special case, namely

LEMMA 3.1.4. *Let D be a division ring satisfying the hypothesis of Theorem 3.1.3. Then D is commutative.*

Proof. Suppose that a, $b \in D$ are such that $c = ab - ba \neq 0$; by hypothesis $c^m = c$ for some $m > 1$. If $\lambda \neq 0 \in Z$, the center of D, then $\lambda c = \lambda(ab - ba) = (\lambda a)b - b(\lambda a)$ hence, by hypothesis, there is an integer $n > 1$ such that $(\lambda c)^n = \lambda c$. Let $q = (m-1)(n-1) + 1$; we see that both $c^q = c$ and $(\lambda c)^q = \lambda c$ hence $(\lambda^q - \lambda)c = 0$. Being in a division ring we deduce that $\lambda^q = \lambda$; since $\lambda^q = \lambda$ for every $\lambda \in Z$, $q > 1$ depending on λ we know from our earlier work that Z is of characteristic $p \neq 0$. Let P be the prime field of Z.

We claim that if D is not commutative we could have chosen our a, b such that not only is $c = ab - ba \neq 0$ but, in fact, c is not even in Z. If not, all commutators are in Z; hence $c \in Z$ and $Z \ni a(ab) - (ab)a = a(ab - ba) = ac$. This would place $a \in Z$ contrary to $c = ab - ba \neq 0$.

Thus we may assume that $c = ab - ba$ is not in Z. Since $c^m = c$, c is algebraic over P hence $c^{p^k} = c$ for some $k > 0$. All the hypothesis of Lemma 3.1.1 are satisfied for c hence we can find an $x \in D$ such that $xcx^{-1} = c^i \neq c$, that is, $xc = c^i x \neq cx$. In particular $d = xc - cx \neq 0$; but $dc = (xc - cx)c = xc^2 - cxc = c^i(xc - cx) = c^i d$. As a commutator, $d^t = d$ for some $t > 1$ and $dcd^{-1} = c^i$. Thus the multiplicative subgroup of D generated by c and d is finite, hence by Lemma 3.1.2 is abelian. This contradicts $cd \neq dc$ and proves the lemma.

With this lemma established we are able to complete the proof of Theorem 3.1.3.

Let R be a ring in which $(xy - yx)^{n(x,y)} = yx - yx$ for all x, $y \in R$. If R is semisimple it is isomorphic to a subdirect sum of primitive rings R_α each of which, as homomorphic images of R, enjoys the hypothesis placed on R. To show that R is commutative it is enough to prove that the R_α are, in other words we may assume that R is primitive.

In this case either $R \approx D$ for some division ring D—in which case we would deduce that R is commutative by use of Lemma 3.1.4—or for some $k > 1$ D_k is a homomorphic image of a subring of R. We wish to show that this latter possibility does not arise. If it did, D_k as a homomorphic image of a subring of R would inherit the property $(xy - yx)^{n(x,y)} = yx - yx$. This is seen to be patently false by considering the elements

$$x = \begin{pmatrix} 1 & 0 & \cdots & 0 \\ 0 & 0 & \cdots & 0 \\ . & . & & . \\ 0 & 0 & \cdots & 0 \end{pmatrix}, \quad y = \begin{pmatrix} 0 & 1 & 0 & \cdots & 0 \\ 0 & 0 & 0 & \cdots & 0 \\ . & . & . & & . \\ 0 & 0 & 0 & \cdots & 0 \end{pmatrix}$$

for these satisfy $0 \neq y = xy - yx$ and $y^2 = 0$. Thus, if R is semisimple it must be commutative.

Suppose now that R is any ring satisfying the hypothesis $(xy - yx)^{n(x,y)} = (xy - yx)$; then $R/J(R)$ is a semisimple ring enjoying the same hypothesis, so by the above discussion it must be commutative. Therefore, given x, $y \in R$ then $xy - yx \in J(R)$. In that case, since $xy - yx \in J(R)$ and $(xy - yx)^{n(x,y)} = (xy - yx)$ we conclude (as we have in similar situations before) that $xy - yx = 0$. In other words, we have shown R to be commutative.

The result can easily be extended to higher commutators as follows: define $[x_1, \cdots, x_m]$ inductively by $[x_1, x_2] = x_1 x_2 - x_2 x_1$, $[x_1, \cdots, x_m] = [[x_1, \cdots, x_{n-1}], x_m]$. The obvious adaptation of the argument just given proves: if R is a ring in which $[x_1, \cdots, x_n]^{m(x_1, \cdots, x_n)} = [x_1, \cdots, x_n]$ for all $x_1, \cdots, x_n \in R$ then $[x_1, \cdots, x_n] = 0$ identically in R. Moreover, if R is semisimple (in fact, if R has no nil ideals) it must be commutative.

2. Some special rings.

We extend the results in Section 1 in several directions. These also generalize results obtained by other motivations. We need some well-known field theoretic preliminaries.

Let K and F be fields with K an algebraic extension of F. An element $a \in K$ is said to be separable over F if its minimal polynomial over F has no multiple roots. Since a polynomial $p(x)$ has multiple roots if and only if $p(x)$ and $dp(x)/dx$ have a common factor, we have that an irreducible polynomial has multiple roots only if $dp(x)/dx = 0$. In characteristic 0 this implies that $p(x)$ is a constant, hence every element in K turns out to be separable over F. In characteristic $p \neq 0$ if $dp(x)/dx = 0$ then $p(x) = g(x^p)$. We can find an integer k so that if $a \in K$ then a^{p^k} is separable over F; however it may well happen that $a^{p^k} \in F$. In this case we say that a is purely inseparable over F. The set of inseparable elements over F form a subfield of K. Similarly—but of much greater depth—the elements in K separable over F form a subfield of K. If F_0 is the field of two elements and $F = F_0(x)$ is the rational function field in x over F_0 then it is easy to verify that the field $K = F(y)$ where $y^2 = x$ is a purely inseparable extension of F.

Returning to our questions on commutativity we begin with a field theoretic result due to Kaplansky.

LEMMA 3.2.1. *Let K be an extension field of F, $K \neq F$, and suppose that given $a \in K$ there is an integer $n(a) > 0$ such that $a^{n(a)} \in F$. Then either:*

1. *K is purely inseparable over F or*
2. *K is of prime characteristic and is algebraic over its prime field P.*

Proof. If K is purely inseparable over F there is nothing to prove.

Suppose that K is not purely inseparable over F; hence there is an $a \in K$, $a \notin F$ which is separable over F. Since $a^n \in F$, a is algebraic and separable over F hence the field $F(a)$ can be imbedded in a finite normal extension L of F. The normality of L gives us an automorphism ϕ of L leaving F fixed such that $b = \phi(a) \neq a$. Now $b^n = \phi(a)^n = \phi(a^n) = a^n$ since $a^n \in F$, from which we see that $b = \nu a$ where $\nu \neq 1 \in L$ is an nth root of unity. Similarly, since $\phi(a+1) = b+1$ and $(a+1)^m \in F$ there is a $\mu \in L$, $\mu^m = 1$ with $b+1 = \mu(a+1)$. Now $\nu \neq \mu$ otherwise $b+1 = \nu(a+1) = \nu a + \nu = b + \nu$ contrary to $\nu \neq 1$. Solving for a we get $a = (1-\mu)/(\nu-\mu)$; since μ, ν are roots of unity they are algebraic over the prime field P whence a is algebraic over P. To finish the proof we must merely show that P is of characteristic $p \neq 0$.

Let L_0 be a finite normal extension of P containing a. The argument used above for a also works for $a+i$, i any integer (for if a is separable in K over F so is $a+i$), leading to $a+i = (i-\mu_i)/\nu_i-\mu_i$) where the ν_i, μ_i are roots of unity and lie in L_0. If the characteristic is 0 then all the $a+i$ are distinct hence L_0, a finite extension of the rationals, would have an infinite number of distinct roots of unity. This is impossible, hence the characteristic is $p \neq 0$.

If $f \in F$ then $a+f$ is also separable over F hence $a+f$ is algebraic over P; since a is algebraic over P we deduce that f is. So F is algebraic over P; but K is algebraic over F. In short K is algebraic over P. This proves the lemma.

In studying division algebras the need for separable elements makes itself felt quite often. Fortunately their existence is assured in fairly general circumstances. The result we are about to prove was first established by Emmy Noether for finite dimensional division algebras and extended by Jacobson to algebraic ones.

THEOREM 3.2.1. *If D is a noncommutative division algebra which is algebraic over its center Z then there is an element in D, not in Z, which is separable over Z.*

Proof. If D is of characteristic 0 there is nothing to prove since every element in D is separable over Z. Hence we consider a division ring of characteristic $p \neq 0$. If the theorem were false then D would be purely inseparable over Z, that is, given $x \in D$ then $x^{p^{n(x)}} \in Z$ for some $n(x) \geq 0$. Therefore there is an $a \in D$, $a \notin Z$ such that $a^p \in Z$. Let δ be defined on D by $x\delta = xa - ax$; as we are in characteristic $p \neq 0$, $x\delta^p = xa^p - a^p x = 0$ since $a^p \in Z$. Since $a \notin Z$, $\delta \neq 0$. Thus if $y\delta \neq 0$ there is a $k > 1$ such that $y\delta^k = 0$ but $y\delta^{k-1} \neq 0$. Let $x = y\delta^{k-1}$; then from $k > 1$, $x = w\delta = wa - aw$. Also from $x\delta = 0$, $xa = ax$. Moreover, being in a division ring we can write $x = au$; since x commutes with a so does u. Thus $au = wa - aw$ giving us $a = (wa - aw)u^{-1} = (wu^{-1})a - a(wu^{-1}) = ca - ac$ where $c = wu^{-1}$. This in turn yields $c = 1 + aca^{-1}$. But for some

t, $c^{p^t} \in Z$ hence $c^{p^t} = (1 + aca^{-1})^{p^t} = 1 + (aca^{-1})^{p^t}$ $= 1 + ac^{p^t}a^{-1} = 1 + c^{p^t}$ (since $c^{p^t} \in Z$). This gives the absurd conclusion $1 = 0$.

The preceding two results enable us to get a theorem which is both a generalization of Jacobson's Theorem (Theorem 3.1.2) and the Jacobson-Noether Theorem (Theorem 3.2.1). It was first proved for semisimple rings by Kaplansky and, in the form given here, by us.

THEOREM 3.2.2. *Let R be a ring with center Z and, given $a \in R$, suppose there exists an integer $n(a) > 0$ such that $a^{n(a)} \in Z$. Then if R has no nil ideals it must be commutative. Equivalently, the commutator ideal of R must be nil.*

Proof. We first derive the result for division rings. If R is a division ring, since it is algebraic over Z, by Theorem 3.2.1 either R is commutative or it has an element $a \notin Z$ which is separable over Z. In this latter possibility the field $Z(a)$ is not purely inseparable over Z and satisfies the hypothesis of Lemma 3.2.1. The outcome of this is that $Z(a)$, and hence Z, is algebraic over the prime field P which is of characteristic $p \neq 0$. Given $x \in R$ it is algebraic over Z hence is algebraic over P, which is to say that $P(x)$ is a finite field. Thus $x^{m(x)} = x$ for some $m(x) > 1$; by Jacobson's Theorem R must be commutative.

Having obtained the result for division rings it is simple to get it for primitive rings. If R is primitive either it is a division ring D or D_k for some $k > 1$ is a homomorphic image of a subring of R. But in D_k the element

$$e = \begin{pmatrix} 1 & 0 & \cdots & 0 \\ 0 & 0 & \cdots & 0 \\ \cdot & \cdot & & \cdot \\ 0 & 0 & \cdots & 0 \end{pmatrix}$$

satisfies $e^m = e$ for all m and is not in the center of D_k. We conclude that R must be isomorphic to D hence must be commutative.

We now make a switch on the general procedure outlined earlier. True, what we have done so far gives the result for semi-simple rings, but we want it in a wider setting. So we seek another path.

Suppose that R is a ring with no nil ideals which satisfies the condition $a^{n(a)} \in Z$. By Lemma 2.2.3 and the remarks made right after its proof we can represent R as a subdirect sum of prime rings R_α which enjoy the following further property: there is a nonnilpotent element $x_\alpha \in R_\alpha$ such that given a nonzero ideal $U_\alpha \subset R_\alpha$ then $x_\alpha^{m(U)} \in U_\alpha$. As homomorphic images of R the R_α satisfy the hypothesis $a^{n(a)} \in Z_\alpha$. Thus to prove the theorem it suffices to prove it for the special case of the R_α.

Hence we assume that R is a prime ring satisfying $a^{n(a)} \in Z$ and furthermore there is a nonnilpotent element $b \in R$ such that given a nonzero ideal U of R then $b^{m(U)} \in U$. Since $b^{n(b)} = c \in Z$ is also not nilpotent and its powers sweep out the nonzero ideals of R, we may *ab initio* assume that $b \in Z$. Since R is prime no element of Z is a zero divisor in R.

Let $\mathcal{R} = \{(r, z) \mid r \in R;\ z \neq 0 \in Z\}$; in $\mathcal{R}$ we define an equivalence relation by $(r_1, z_1) \smile (r_2, z_2)$ if $r_1 z_2 = r_2 z_1$. This indeed is an equivalence relation; let R^* be the set of equivalence classes; denote by $[r, z]$ the class of (r, z). We define $[r_1, z_1] + [r_2, z_2] = [r_1 z_2 + r_2 z_1, z_1 z_2]$ and $[r_1, z_1][r_2, z_2] = [r_1 r_2, z_1 z_2]$. Since the elements of Z are not zero divisors in R we find that these operations are well defined and R^* is a ring. Moreover the mapping $r \rightarrow [rz, z]$ is an imbedding of R in R^*. Finally, the center Z^* of R^* is $\{[r, z] \mid r \in Z\}$; this immediately yields that Z^* is a field.

If $[r, z] \in R^*$ then $[r, z]^{n(r)} = [r^{n(r)}, z^{n(r)}] \in Z^*$, consequently R^* enjoys the same property as does R. However R^* is a simple ring. For let $U^* \neq (0)$ be an ideal of R^* and let $U = \{r \in R \mid [r, z] \in U^* \text{ for some } z \in Z\}$. It is immediately verified that $U \neq (0)$ is an ideal of R hence $b^{m(u)} \in U$. Since $0 \neq b^{m(u)} \in Z$ (because b is) we have that U^* contains a nonzero element of Z^*. But Z^* being a field this yields that $U^* = R^*$, hence our assertion that R^* is simple has been substantiated. As a simple ring with unit element R^* is primitive so must be commutative; this from the first part of our proof. Since $R \subset R^*$ we end up with R commutative. This proves the theorem.

Note that what we did in the proof was imbed R in $R^* = R \otimes_Z Z^*$ where Z^* is the ring of quotients of Z and exploit the nature of R to pin down R^* as simple.

Note that this theorem does generalize the one due to Jacobson which we proved earlier. For let R be a ring in which $x^{n(x)} = x$; then R has no nilpotent elements. If e is an idempotent in R then for all $x \in R$, $(xe - exe)^2 = (ex - exe)^2 = 0$ leaving us with $xe - exe = ex - exe = 0$. Hence $xe = ex$ and all idempotents are in the center of R. Now if $a^{n(a)} = a$ then $e = a^{n(a)-1}$ is an idempotent, hence $a^{n(a)-1} \in Z$. Since R has no nilpotent elements it has no nil ideals, whence the hypothesis of the last theorem is satisfied and R results commutative.

We close this chapter with another commutativity theorem. The theorem we prove is but a special case of much more general results, however, to prove these more general results would require a rather wide digression into field theory. To see these results refer to [5], [6], [11], [21]. We shall give some of their statements after we prove the theorem.

THEOREM 3.2.3. *Let R be a ring with center Z such that for a fixed integer $n > 1$, $x^n - x \in Z$ for all $x \in R$. Then R is commutative.*

The proof will be broken down into a sequence of lemmas. In what follows, unless otherwise stated, R will be a ring in which $x^n - x \in Z$ for all $x \in R$.

LEMMA 3.2.2. *If R is semisimple it is commutative.*

Proof. We begin with the division ring case. Suppose that $x \in R$, $x \notin Z$; if $\lambda \in Z$ then $(\lambda x)^n - \lambda x \in Z$. Together with $x^n - x \in Z$ we arrive at $(\lambda^n - \lambda)x \in Z$ for all $\lambda \in Z$. Since $x \notin Z$ we conclude that $\lambda^n = \lambda$ for all $\lambda \in Z$. Hence Z is a finite field. Since R is algebraic over Z we get $x^{n(x)} = x$ for all $x \in R$ implying, via Jacobson's theorem, the commutativity of R.

If R is primitive and not a division ring then D_k for some $k > 1$ and some division ring D is a homomorphic image of a subring of R. Hence in D_k the condition $x^n - x \in Z$ carries over. But if

$$x = \begin{pmatrix} 0 & 1 & 0 & \cdots & 0 \\ 0 & 0 & 0 & \cdots & 0 \\ . & . & . & & . \\ 0 & 0 & 0 & \cdots & 0 \end{pmatrix}$$

then $x^2 = 0$ hence $x^n - x = -x \in Z$, a contradiction. Thus R is a division ring, hence a field. Since a semisimple ring R is a subdirect sum of primitive rings R_α which are homomorphic images of R and since by the above discussion the R_α are commutative we get that R itself is commutative.

For a general R satisfying the hypothesis, since $R/J(R)$ is semisimple, we have the

COROLLARY. *For all x, $y \in R$, $xy - yx \in J(R)$.*

LEMMA 3.2.3. $J(R) \subset Z$.

Proof. As we have seen, if $\lambda \in Z$ and $x \in R$ then $(\lambda^n - \lambda)x \in Z$, hence for all $y \in R$ $(\lambda^n - \lambda)(xy - yx) = 0$.

If $\lambda \in Z \cap J(R)$ then $\lambda^{n-1} \in J(R)$ and so $(1-\lambda^{n-1})t = 0$ implies that $t = 0$. Since $(1-\lambda^{n-1})\lambda(xy-yx) = 0$ we get that $\lambda(xy-yx) = 0$ for all $\lambda \in Z \cap J(R)$ and all $x, y \in R$. Now if $a \in J(R)$ then $a^n - a \in Z \cap J(R)$ so, by the above, $(a^n-a)(xy-yx) = 0$. However, as before, from $(1-a^{n-1})a(xy-yx) = 0$ with $a \in J(R)$ we get $a(xy-yx) = 0$. Similarly we get that $(xy-yx)a = 0$ for all $a \in J(R)$ and $x, y \in R$.

Putting $x = a$ in these relations yields that $a^2y = aya = ya^2$ for all $a \in J(R)$, $y \in R$ which is to say that $a^2 \in Z(R)$. If n is even then $a^n \in Z$ follows and so, from $a^n - a \in Z$, we get that $a \in Z$. On the other hand if n is odd then $a^{n-1} \in Z$ hence from $a^n - a \in Z$ we have that for all $x \in R$, $0 = (a^n-a)x - x(a^n-a) = (a^{n-1}-1)(xa-ax)$. Since $a^{n-1} \in J(R)$ this forces $xa-ax = 0$, that is, $a \in Z$. Hence in all cases $a \in Z$, that is $J(R) \subset Z$.

In the course of the proof we saw that $J(R)(xy-yx) = (0)$; now by the corollary to Lemma 3.2.2 $xy-yx \in J(R)$. Thus $(xy-yx)^2 = 0$. Also since $n > 1$, $(xy-yx)^n = 0$ hence from $(xy-yx)^n - (xy-yx) \in Z$ we deduce that $xy-yx \in Z$. We summarize this in the

COROLLARY. *For all* x, $y \in R$, $(xy-yx)^2 = 0$ *and* $xy-yx \in Z$.

Let us recall that in Section 2 of Chapter 2 we introduced the concept of subdirectly irreducible ring; a ring is said to be subdirectly irreducible if the intersection of all its nonzero ideals is not (0). In Lemma 2.2.2 we showed that any ring is a subdirect sum of subdirectly irreducible rings. Hence in order to prove our theorem it is sufficient to do it for subdirectly irreducible rings.

We henceforth assume in the proof that R *is a subdirectly irreducible ring* in which $x^n - x \in Z$ for all $x \in R$. Let $S \neq (0)$ be the intersection of the nonzero ideals of R. Of course S is the unique minimal ideal of R. By

Lemma 3.2.2 we may assume that $J(R) \neq (0)$, otherwise we know R to be commutative. Thus $S \subset J(R)$; since by Lemma 3.2.3 $J(R) \subset Z$ we know that $S \subset Z$. Since we have shown that $J(R)(xy-yx) = (0)$ we readily obtain from this that if R is not commutative then $S^2 = (0)$.

We now prove

LEMMA 3.2.4. *There exists a prime p such that $p(xy-yx) = 0$ for all $x, y \in R$.*

Proof. Since $x^n - x \in Z$ and $(2x)^n - 2x \in Z$ we get $(2^n - 2)x \in Z$ hence $(2^n - 2)(xy - yx) = 0$. If R is not commutative it thus has elements of finite additive order, hence of prime order p. Let $R_p = \{x \in R \mid px = 0\}$; $R_p \neq (0)$ is clearly an ideal of R hence $R_p \supset S$. If $R_p \neq (0)$ for some prime $q \neq p$ then $R_q \supset S$. Since $S \subset R_q \cap R_p = (0)$ we would have a contradiction.

Now $(p^n - p)(xy - yx) = 0$, that is $(p^{n-1} - 1)p(xy - yx) = 0$. Since $p^{n-1} - 1$ is relatively prime to p, by the above remark we conclude that $p(xy - yx) = 0$ for all $x, y \in R$.

Let $x, y \in R$; by the corollary to Lemma 3.2.3 $xy - yx \in Z$. What is $x^2 y - yx^2$? Calculating we get $x^2 y - yx^2 = x(xy - yx) + (xy - yx)x = 2x(xy - yx)$. Continuing in this way we arrive at $x^k y - yx^k = kx^{k-1}(xy - yx)$. In particular, putting $k = p$ we have $x^p y - yx^p = px^{p-1}(xy - yx) = 0$. We have proved

LEMMA 3.2.5. *For all $x \in R$, $x^p \in Z$.*

Let $A(S) = \{x \in R \mid xS = (0)\}$; $A(S)$ is an ideal of R and since $S^2 = (0)$, $S \subset A(S)$ hence $A(S) \neq (0)$.

Let $x \in A(S)$; by Lemma 3.2.5 $x^p \in Z$ therefore $(x^{np} - x^p)(yz - zy) = 0$ for all $y, z \in R$. Thus

$$x^{(n-1)p} x^p (yz - zy) = x^p (yz - zy).$$

Let $T = \{r \in R \mid x^{(n-1)p} r = r\}$; because $x^{(n-1)p} \in Z$, T turns out to be an ideal of R. Therefore if $T \neq (0)$ we would have $T \supset S$. But this is false for if $0 \neq r \in S$ then, since

$x \in A(S)$, $x^{(n-1)p}r = 0 \neq r$. Hence $T = (0)$. But $x^p(yz - zy)$ $\in T = (0)$ so we get that $x^p(yz - zy) = 0$ for all $x \in A(S)$ and all y, $z \in R$.

In particular this yields that $x^p(xy - yx) = 0$, that is, $x^{p+1}y = x^p yx = yx^{p+1}$ since $x^p \in Z$. Continuing we obtain that $x^{p+k}y = yx^{p+k}$ for all $k \geq 0$, and so $x^{p+k} \in Z$.

Now from $x^n - x \in Z$ and $(x^n)^n - x^n \in Z$ we derive $x^{n^2} - x \in Z$; in fact we can go on to get $x^{n^k} - x \in Z$ for all $k \geq 1$. Pick k so that $n^k > p$; by the argument above for $x \in A(S)$, since $n^k > p$, $x^{n^k} \in Z$. Thus from $x^{n^k} \in Z$ and $x^{n^k} - x \in Z$ we end up with $x \in Z$. We have proved

LEMMA 3.2.6. $A(S) \subset Z$.

This lemma assures us that if, in particular, $A(S) = R$ then R is commutative. *We assume in the sequel that* $A(S) \neq R$.

Suppose that $a \neq 0$ is a zero divisor in R, that is, $ax = 0$ for some $x \neq 0$; then certainly $ax^p = 0$. If $x^p = 0$ then since $x^{n^k} - x \in Z$ and $n^p > p$ we would have $x \in Z$. If $x^p \neq 0$ then it is a nonzero element in Z by Lemma 3.2.5. At any rate we get that a annihilates a nonzero center element z. Since $z \neq 0 \in Z$ we claim that $Rz \neq (0)$ for otherwise $\{x \in R \mid Rx = (0)\}$ is a nonzero ideal of R and so contains S; but this would yield $RS = (0)$ hence $R = A(S)$ contrary to assumption. Thus $Rz \neq (0)$; because $z \in Z$, Rz is an ideal of R hence $Rz \supset S$. Thus $(0) = azR = aRz \supset aS$, which is to say that $a \in A(S)$. We have proved

LEMMA 3.2.7. *All the zero divisors of R are in $A(S)$.*

In order to complete the proof of Theorem 3.2.3 we need one more basic lemma.

LEMMA 3.2.8. *$R/A(S)$ is a finite field.*

Proof. We first establish that $R/A(S)$ is a field. If $s \neq 0$ is in S, as an element of Z we have Rs is an ideal of

R. As we saw earlier, if $Rs = (0)$ then we would end up with $R = A(S)$, contrary to assumption. As a nonzero ideal $Rs \supset S$; since $s \in S$, $Rs \subset S$ hence $Rs = S$.

If $x \in R$, $x \notin A(S)$, then by Lemma 3.2.7, x is not a zero divisor, hence $xs \neq 0 \in S$. By the above, $Rxs = S$; if $z \in R$, $z \notin A(S)$ then also $Rzs = S$. Hence there is a $y \in R$ such that $yxs = zs$ and so $(yx - z)s = 0$. As a zero divisor $yx - z \in A(S)$. This says that in $\overline{R} = R/A(S)$ we can solve the equation $\bar{y}\bar{x} = \bar{z}$ for $\bar{y}$ if $\bar{x} \neq 0$ and $\bar{z}$ are given. Hence $R/A(S)$ is a division ring. Since all $xy - yx$ are zero divisors they are in $A(S)$, hence $R/A(S)$ is commutative. So far we have proved that $R/A(S)$ is a field.

Now to the finiteness of $R/A(S)$. If $x \notin A(S)$ then since $x^p \in Z$, $(x^{np} - x^p)(yz - zy) = 0$ for all y, $z \in R$. Thus if R is not commutative, $x^{np} - x^p$ is a zero divisor so is in $A(S)$. Hence in the field $\overline{R} = R/A(S)$ every element satisfies the fixed polynomial $\bar{x}^{np} = \bar{x}^p$. This forces $\overline{R}$ to be finite.

We now have all the ingredients to complete the proof of Theorem 3.2.3.

Since $\overline{R} = R/A(S)$ is a finite field its multiplicative group of nonzero elements is cyclic. Let $\bar{a}$ be a generator of this group and let $a \in R$ map on $\bar{a}$. If $x \notin A(S)$ then $a^t - x \in A(S) \subset Z$ for some t, hence $(a^t - x)a = a(a^t - x)$ giving us $ax = xa$ for all $x \notin A(S)$. Since $A(S) \subset Z$, certainly $ax = xa$ for all $x \in A(S)$. All in all, we have proved that $a \in Z$. From $a^t - x \in A(S) \subset Z$, for $x \notin A(S)$, we deduce that $x \in Z$. Since both $A(S)$ and its complement are in Z we have shown that $R = Z$. The theorem is now proved.

There are various generalizations of this theorem—to see their proofs consult the papers in the bibliography that follows. For instance one can weaken the condition that n be fixed to allow n to be a function of x. In fact one can just assume the existence of a polynomial $p_x(t)$

with integer coefficients depending on x such that $x^2 p_x(x) - x \in Z$; this can be shown to force the commutativity of R. One can give local versions of these theorems, for instance, suppose that given x, $y \in R$ there exists an integer $n = n(x, y)$, such that $x^n - x$ commutes with y; then R is commutative. The other theorems also have their localized analogs.

Another situation which is close to commutativity is when the commutator ideal of R is nil. We saw an instance of this in Theorem 3.2.2. There are many other natural hypotheses which render nil the commutator ideal of a ring. For instance if R is a ring in which $(xy)^n = x^n y^n$, $n > 1$ a fixed integer then R must have a nil commutator ideal. To cite still another, if R is a ring in which given x and y then there are integers $n(x, y)$, $m(x, y) > 0$ such that $(xy)^{n(x,y)} = (yx)^{m(x,y)}$ then again the commutator ideal of R is nil.

References

1. E. Artin, Über einen Satz von Herrn J. H. MacLaglan-Wedderburn, *Abh. Math. Seminar*, Univ. Hamburg, 5 (1927) 245–250.

2. R. Belluce, S. K. Jain and I. N. Herstein, Generalized commutative rings, *Nagoya Math. J.*, 27 (1966) 1–5.

3. C. C. Faith, Radical extensions of rings, *Proc. Amer. Math. Soc.*, 12 (1961) 274–283.

4. I. N. Herstein, A generalization of a theorem of Jacobson, I, *Amer. J. Math.*, 73 (1951) 755–762.

5. ———, A generalization of a theorem of Jacobson, III, *Amer. J. Math.*, 75 (1953) 105–111.

6. ———, The structure of a certain class of rings, *Amer. J. Math.*, 75 (1953) 864–871.

7. ———, An elementary proof of a theorem of Jacobson, *Duke Math. J.*, 21 (1954) 45–48.

8. ———, A theorem on rings, *Canad. J. Math.*, 5 (1953) 238–241.

9. ———, A theorem concerning three fields, *Canad. J. Math.*, 7 (1955) 202–203.

10. ———, Wedderburn's theorem and a theorem of Jacobson, *Amer. Math. Monthly*, 68 (1961) 249–251.

11. O. N. Herstein, Two remarks on the commutativity of rings, *Canad. J. Math.*, 7 (1955) 411–412.

12. ———, A condition for the commutativity of rings, *Canad. J. Math.*, 9 (1957) 583–586.

13. ———, Power maps in rings, *Michigan Math. J.*, 8 (1961) 29–32.

14. M. Ikeda, On a theorem of Kaplansky, *Osaka Math. J.*, 4 (1952) 235–240.

15. N. Jacobson, *Structure of rings*, Amer. Math. Soc. Colloq. Publ., 37 (1964)

16. ———, Structure theory for algebraic algebras of bounded degree, *Ann. of Math.*, 46 (1945) 695–707.

17. I. Kaplansky, A theorem on division rings, *Canad. J. Math.*, 3 (1951) 290–292.

18. W. Martindale, The structure of a certain class of rings, *Proc. Amer. Math. Soc.*, 9 (1958) 714–721.

19. ———, The commutativity of a certain class of rings, *Canad. J. Math.*, 12 (1960) 263–268.

20. M. Nagata, T. Nakayama and T. Tuzuku, On an existence lemma in valuation theory, *Nagoya Math. J.*, 6 (1953) 59–61.

21. T. Nakayama, Über die Kommutativität gewisser Ringe, *Hamb. Abhand*, 20 (1955) 20–27.

22. ———, A remark on the commutativity of algebraic rings, *Nagoya Math. J.*, 14 (1959) 39–44.

23. J. H. M. Wedderburn, A theorem on finite algebras, *Trans. Amer. Math. Soc.*, 6 (1905) 349–352.

CHAPTER 4

SIMPLE ALGEBRAS

Wedderburn's pioneer work on the structure of simple algebras set the stage for deep investigations—often with an eye to application in algebraic number theory—in the theory of algebras. Much of the early research, following on the heels of that of Wedderburn, came in the work of Dickson. Then in the 1920's and early 1930's a very deep investigation of simple algebras was carried out culminating in a beautiful structure theory for division algebras over algebraic number fields. A large part of the results was developed in the hands of Albert, Artin, Brauer, Noether and many others.

As is often the case in mathematics, the cycle swings back on itself. This lovely work on the theory of simple algebras has served as the inspiration for much work in the algebra of today. Now, in the setting of homological algebra, these results on simple algebras have been smoothed, restated, reinterpreted and extended to an extremely large setting. To see the particulars of this activity one should consult some of the papers we cite at the end, particularly those of Amitsur, Auslander, Chase, Harrison, Rosenberg, Serre and Zelinsky.

1. The Brauer group. We begin with a

DEFINITION. *An algebra A is said to be central simple over a field F if A is a simple algebra having F as its center.*

Our concern shall be the nature of the set of simple algebras central over a fixed field F. Thus it is desirable to know what we can do to such algebras and still stay in the set. A step in this direction and an important one—albeit a simple one—is provided in

LEMMA 4.1.1. *If A is central simple over F and B is a simple algebra containing F in its center then $A \otimes_F B$ is simple.*

Proof. Let $U \neq (0)$ be an ideal of $A \otimes_F B$. If $u \neq 0 \in U$ write u as $u = \sum a_i \otimes b_i$ where $a_i \in A$, $b_i \in B$ and where the b_i are linearly independent over F. Let us call the number of nonzero a_i's in this expression for u the length of u. Pick $u \neq 0$ in U of shortest length. If $r, s \in A$ then $(r \otimes 1) u (s \otimes 1) = \sum r a_i s \otimes b_i$ is in U. Since A is simple $A a_i A = A$ therefore we can arrange it so that there is an element u_1 in U of the same length as u but of the form $u_1 = 1 \otimes b_1 + a_2' \otimes b_2 + \cdots + a_m' \otimes b_m$. Given $a \in A$ then $(a \otimes 1) u_1 - u_1 (a \otimes 1)$ is in U; writing it out explicitly we get $(a_2' a - a a_2') \otimes b_2 + \cdots + (a_m' a - a a_m') \otimes b_m$ is in U. However, this is of shorter length than u_1 so must be 0. Since the b_i are linearly independent over F from the properties of the tensor product the $1 \otimes b_i$ are linearly independent over $A \otimes 1$, the net result of which is that $a_i' a = a a_i'$ for $i = 2, \cdots, m$ for all $a \in A$. In short each a_i' is in the center of A so must be in F. We write $a_i' = \alpha_i \in F$. Thus $u_1 = 1 \otimes b_1 + \alpha_2 \otimes b_2 + \cdots + \alpha_m \otimes b_m$ $= 1 \otimes (b_1 + \alpha_2 b_2 + \cdots + \alpha_m b_m)$ is in U. Since the b_i are independent over F, $b = b_1 + \alpha_2 b_2 + \cdots + \alpha_m b_m \neq 0$; hence $U \supset (1 \otimes B) u_1 (1 \otimes B) = 1 \otimes B b B = 1 \otimes B$ from the simplicity of B. From this we get that $(A \otimes 1)(1 \otimes B) \subset U$; but $(A \otimes 1)(1 \otimes B) = A \otimes B$. This yields that $U = A \otimes B$, in other words $A \otimes B$ is simple.

Note that the above argument could easily be modified to study the ideal structure of $A \otimes B$ even when B is not simple. Since we shall have no need of this in the sequel we leave that to the reader.

If A and B are both central simple over F by the lemma we know, of course, that $A \otimes_F B$ is simple. However, here we can explicitly calculate its center. We do this in

THEOREM 4.1.1. *If A and B are central simple over F then $A \otimes_F B$ is central simple over F.*

Proof. To prove the theorem, since we already know that $A \otimes_F B$ is simple, we must merely demonstrate that its center is F.

Suppose that $z = \sum a_i \otimes b_i$ is in the center of $A \otimes B$; here again we assume that the b_i are linearly independent over F. Given $a \in A$ then $0 = (a \otimes 1)z - z(a \otimes 1) = \sum (aa_i - a_i a) \otimes b_i$; this yields that each $aa_i - a_i a = 0$ for all $a \in A$. Therefore each a_i is in the center of A, hence in F. If we write $a_i = \alpha_i \in F$ then we have that $z = \sum a_i \otimes b_i = 1 \otimes \sum \alpha_i b_i = 1 \otimes b$. Since z is in the center of $A \otimes B$, for any $x \in B$, $z(1 \otimes x) - (1 \otimes x)z = 0$; calculating this explicitly yields that b is in the center of B, hence in F. In other words $z = \beta(1 \otimes 1)$ where $\beta \in F$. We see that the center of $A \in B$ is precisely $F(1 \otimes 1)$, that is, F if we identify F with $F(1 \otimes 1)$. The algebra $A \otimes_F B$ has been shown to be central simple over F.

These last results have a very interesting consequence for division algebras finite-dimensional over their centers.

THEOREM 4.1.2. *If D is a finite-dimensional division algebra over its center Z then the dimension of D over Z is a perfect square.*

Proof. Let $\overline{Z}$ be the algebra closure of Z. Since D is certainly central simple over Z invoking Lemma 4.1.1 we have that $\overline{D} = D \otimes_Z \overline{Z}$ is simple. Moreover $[D : Z] = [\overline{D} : \overline{Z}]$. Now, by Wedderburn's Theorem, since $\overline{D}$ is simple and finite-dimensional over the algebraically closed field $\overline{Z}$, $\overline{D} \approx \overline{Z}_n$. Consequently $[\overline{D} : \overline{Z}] = n^2$; from the equality of $[D : Z]$ and $[\overline{D} : \overline{Z}]$ we conclude that $[D : Z] = n^2$.

If A is simple and finite-dimensional over its center Z, by Wedderburn's Theorem $A \approx D_m$ where D is a finite-dimensional division algebra with Z as its center. By the theorem above $[D:Z]=n^2$ hence $[A:Z]=(nm)^2$. Therefore we have the

COROLLARY. *If A is a simple algebra finite-dimensional over its center Z then $[A:Z]$ is a perfect square.*

If R is any ring let R^* be the ring whose additive group is that of R but where the product is defined by $a \cdot b = ba$—that is, $\cdot$ is the product in R^*, ba is that in R. Clearly R^* is anti-isomorphic to R. We call R^* the *reverse* of R.

We prove the basic

THEOREM 4.1.3. *If A is a simple algebra finite-dimensional over its center F then $A \otimes_F A^* \approx F_n$ where $n = \dim {}_F A$.*

Proof. Considering A merely as a vector space over F then the ring of linear transformations on A over F, $L(A)$, is of dimension n^2 over F, and is isomorphic to F_n.

Let $A_r = \{T_a \mid a \in A\}$ where $T_a \in L(A)$ maps any x onto xa. Let $A_l = \{L_a \mid a \in A\}$ where $L_a \in L(A)$ maps any x onto ax. Clearly A_r is isomorphic to A and A_l is isomorphic to A^*. Any element in A_r commutes with any element in A_l. From the isomorphism stated above, $A \otimes A^* \approx A_r \otimes A_l$. Since A_r and A_l are central simple over F so is $A_r \otimes A_l$. We map $A_r \otimes A_l$ into $A_r A_l \subset L(A)$ by the mapping $\sum T_a \otimes L_b \to \sum T_a L_b$. Trivially this mapping is onto $A_r A_l$; from the commuting properties of A_r with A_l this mapping is a homomorphism. The simplicity of $A_r \otimes A_l$ forces this to be an isomorphism hence we have shown that $A \otimes_F A^* \approx A_r A_l$. But since

$$n^2 = \dim {}_F L(A) \geq \dim {}_F A_r A_l = \dim {}_F (A \otimes A^*)$$
$$= (\dim {}_F A)^2 = n^2$$

we get that $\dim {}_F A_r A_l = \dim {}_F L(A)$. Since $A_r A_l$ is a subspace of $L(A)$ we conclude that $A_r A_l = L(A)$; hence $A \otimes_F A^* \approx A_r A_l = L(A)$. This is the assertion of the theorem.

If Q is the ring of quaternion over the real field F then Q has a mapping $x \to \bar{x}$ which is an anti-isomorphism of period 2. Hence for Q^* we can take Q itself; applying the theorem we see that $Q \otimes_F Q^* = Q \otimes_F Q \approx F_4$, the 4×4 matrices over the real field.

In the results developed to this point we have the necessary ingredients for defining the *Brauer group* of the field F. To do so we must introduce an appropriate equivalence relation.

DEFINITION. *If A and B are finite-dimensional central simple algebras over the field F then $A \sim B$ if for some integers m and n, $A \otimes_F F_m \approx B \otimes_F F_n$.*

Another way of looking at this equivalence is the following: by Wedderburn's theorem $A \approx D_1 \otimes F_n$ and $B \approx D_2 \otimes F_m$ for some finite dimensional division algebras having F as center. Then $A \sim B$ if and only if $D_1 \approx D_2$. In other words, the equivalence relation defined for central simple algebras over F is, in the final count, one really defined for central division algebras over F. We keep referring to this relation as an equivalence relation; we leave to the reader the verification that it is indeed so.

Let $B(F)$ denote the equivalence classes of finite-dimensional central simple algebras over F. If $[A]$ denotes the class of A we define a product in $B(F)$ by $[A][B] = [A \otimes_F B]$. Theorem 4.1.1 assures us that $B(F)$ is closed under this product. From the properties of the tensor product this product in $B(F)$ is associative and commutative. If $[F]$ is the equivalence class of F (note that $[A] = [F]$ if and only if $A \approx F_n$ for some n) the $[F]$ acts as a unit element for $B(F)$. Finally, for $[A] \in B(F)$ if $[A^*]$ is the class of A^*, the reverse of A, then, by Theorem 4.1.3, $[A][A^*] = [F]$. In summary, we have verified

THEOREM 4.1.4. $B(F)$ *is an abelian group.*

$B(F)$ is called the *Brauer group* of F. Since for any A, $[A] = [D]$ for a unique (up to isomorphism) division algebra D, this invariant $B(F)$ of the field F measures the division algebras lying above F and having F as its center. We shall return to a more thorough study of the nature of $B(F)$ later.

2. Maximal subfields. The maximal subfields of a division ring influence in a very strong way the global nature of the division ring itself. There are some very beautiful, classical theorems that try to describe the way in which these maximal subfields sit in the division ring. It is some of these that we shall develop now.

If D is a division ring and $S \subset D$ then the *centralizer* of S in D, $C(S)$, is $\{x \in D \mid xs = sx$ for all $s \in S\}$. It is trivially a subdivision ring of D.

A subfield K of D is, naturally enough, called a *maximal subfield* if it is not properly contained in a larger subfield of D. Note that K must then automatically contain Z for otherwise $Z(K)$, the field obtained by adjoining the elements of K to Z, would be strictly larger than K.

LEMMA 4.2.1. *If D is a division ring and K is a subfield of D then K is a maximal subfield of D if and only if $K = C(K)$.*

Proof. If $K = C(K)$ and $L \supset K$ is a subfield of D then since $L \subset C(K) = K$ we get $L = K$ and so K is a maximal subfield.

If K is a maximal subfield and $a \in C(K)$ then $K(a) \supset K$ is a subfield of D which contains K. In consequence,

$K(a) = K$ and so $a \in K$. Thus $C(K) \subset K \subset C(K)$ yielding that $K = C(K)$.

We have stated in the first few words of this section that the nature of the maximal subfields is crucial in determining that of the division ring. An indication of this interplay is given in the next result.

THEOREM 4.2.1. *Let D be a division ring with center F and let K be a maximal subfield of D; then $D \otimes_F K$ is a dense ring of linear transformations on D considered as a vector space over K.*

Proof. In $E(D)$, the ring of endomorphisms of the additive group of D, let $D_r = \{T_a \mid a \in D,\ T_a : x \to xa$ for $x \in D\}$ and $K_l = \{L_k \mid k \in K,\ L_k : x \to kx$ for $x \in D\}$. Clearly D_r and K_l are in $E(D)$ and any element in K_l commutes with any element of D_r.

Since D is a division ring, for $d \neq 0 \in D$, $dD_r = D$ hence D_r acts irreducibly on D; consequently $D_r K_l$ acts irreducibly on D. As a ring of endomorphisms on D it acts faithfully on D.

Let Δ be the commuting ring of $D_r K_l$ on D, that is, Δ is the centralizer of $D_r K_l$ in $E(D)$. Since Δ centralizes D_r it must be contained in $D_l = \{L_a \mid a \in D\}$. Since Δ also centralizes K_l, which is a maximal subfield of D_l, we get that $\Delta \subset K_l$. However, K_l clearly contains Δ hence $\Delta = K_l$. We have shown that $D_r K_l$ is a dense ring of linear transformations on D over K_l.

Now since K is commutative, K is isomorphic to K_l. By Lemma 4.1.1 $D \otimes_F K$ is simple; the mapping $D \otimes_F K \to D_r K_l$ defined by $\sum a_i \otimes k_i \to \sum T_{a_i} L_{k_i}$ is a homomorphism onto, hence, is an isomorphism. Thus $D \otimes_F K \approx D_r K_l$ and so is a dense ring of K-linear transformations on D. This proves the theorem.

In the special instance when D is finite-dimensional over its center F the theorem assumes a much sharper form. In that case if K is a maximal subfield then $[D:K]=n<\infty$ and the density of the action of $D\otimes_F K$ on D over K means that $D\otimes_F K\approx K_n$. This is the

COROLLARY. *If D is finite-dimensional over its center F and if K is a maximal subfield of D then $D\otimes_F K\approx K_n$ where $n=[D:K]$.*

Now dim $_K(D\otimes_F K)=$ dim $_F D$; but since $D\otimes_F K\approx K_n$ we have that dim $_K(D\otimes_F K)=n^2$. Thus from $n=[D:K]$ we get, using $[D:K][K:F]=[D:F]=n^2$, that $[K:F]=[D:K]=n$. Although this is implicitly contained in the corollary above, since it is so important we emphasize it as a theorem.

THEOREM 4.2.2. *If D is a division algebra finite-dimensional over its center F then for any maximal subfield K of D we have that $[D:K]=[K:F]=\sqrt{[D:F]}$.*

Note a simple consequence of Theorem 4.2.2. If D is a central, noncommutative division algebra over the real field F then, since the only finite extension of F is the field of complex numbers K and since $[K:F]=2$, we get that $[D:F]=2^2=4$. From this it is easy to show that D must be isomorphic to the ring of quaternions over the reals.

One further comment is in order. If D is a division algebra infinite dimension over a maximal subfield K then $D\otimes_F K$ is not Artinian, for otherwise, by Theorem 4.2.1 we would get that $D\otimes_F K$ would be a matrix algebra over K. Hence, the tensor product, even of very nice Artinian rings, need not be Artinian.

3. Some classic theorems. Before going on to an exposition of two very important and classical theorems of the subject—the Noether-Skolem theorem and the double centralizer theorem—we return for a moment to the nature of the modules of a semisimple Artinian ring.

LEMMA 4.3.1. *If R is a semisimple Artinian ring and M an R-module then M is the direct sum of irreducible R-modules.*

Proof. Since R is a semisimple Artinian ring, $R = R_1 \oplus \cdots \oplus R_k$ where each $R_i = e_i R$, with e_i a central idempotent in R, is a simple Artinian ring. Since $1 = e_1 + \cdots + e_k$ we get that our R-module M is of the form $M = \oplus Me_i$. Now Me_i is a unitary R_i-module and $Me_i R_j = (0)$ for $i \neq j$. Hence it suffices to prove the result in the special case in which R is simple.

In this case, by Wedderburn's Theorem $R \approx D_n$ where D is a division ring. If e_{ii} represents the idempotent matrix

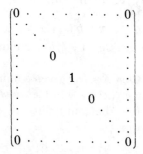

and $\rho_i = e_{ii} R$ then $R = \rho_1 \oplus \cdots \oplus \rho_n$ where each ρ_i, as a minimal right ideal of R, is an irreducible R-module. Now if $m \in M$ then $m = m1 = m(e_{11} + \cdots + e_{nn}) = m_1 + \cdots + m_n$ where $m_i = me_{ii} \in m\rho_i$. Clearly $m\rho_i = (0)$ or $m\rho_i$ is itself an irreducible R-module. Thus $m \in M$ is contained in a sum of irreducible R-modules and so M is the sum of its irreducible submodules. Considering these submodules of M which are direct sums of irreducible R-modules we can apply Zorn's lemma to these to obtain a maximal such, T_0. We assert that $T_0 = M$; if not, let $M_j \not\subset T_0$ for some irreducible submodule M_j of M. Hence by the irreducibility of M_j, $M_j \cap T_0 = (0)$, in which case $M_j + T_0$ is a larger direct sum of irreducible submodules. Thus $T_0 = M$ and the result is established.

Having identified any R-module, for R semisimple and Artinian, as a direct sum of irreducible R-modules we now want to specify which are the irreducible R-modules. This is

LEMMA 4.3.2. *Let R be a semisimple Artinian ring and let M be an irreducible R-module. Then M is isomorphic to a minimal right ideal of R. Moreover, if R is simple, then all its irreducible modules are isomorphic.*

Proof. Let M be an irreducible R-module; by Lemma 1.1.3, M is isomorphic to R/ρ for some maximal right ideal, ρ, of R. By Theorem 1.4.2, $\rho = eR$ for some idempotent $e \in R$. Now $R = eR \oplus (1-e)R$ hence, as an R-module, $M \approx R/\rho = R/eR \approx (1-e)R$, a right ideal of R. Since R/ρ is irreducible, $(1-e)R$ is a minimal right ideal of R.

If R is simple then $R \approx D_n$; given a minimal right ideal ρ it is of the form eR, $e^2 = e \in R$. Hence we can bring e to diagonal form; but the minimality of ρ forces e to be brought to the form

$$\begin{pmatrix} 1 & 0 & \cdots & 0 \\ 0 & 0 & \cdots & 0 \\ \cdot & \cdot & & \cdot \\ 0 & 0 & \cdots & 0 \end{pmatrix}$$

whence we get

$$\rho \approx \begin{pmatrix} \alpha_1 & \cdots & \alpha_n \\ 0 & \cdots & 0 \\ & \cdot & \\ 0 & \cdots & 0 \end{pmatrix}.$$

An elementary theorem in matrix theory asserts that any automorphism of the ring of $n \times n$ matrices over a field leaving the scalars fixed is inner. This theorem has an important extension to simple Artinian rings; this, usually known as the *Noether-Skolem theorem*, is

THEOREM 4.3.1. *Let R be a simple Artinian ring with center F and let A, B be simple subalgebras of R which contain F and are finite-dimensional over it. If ϕ is an isomorphism of A onto B leaving F elementwise fixed then there is an invertible $x \in R$ such that $\phi(x) = x^{-1}ax$ for all $a \in A$.*

Proof. Let $E(R)$, R_e, A_r, B_r have their usual meaning (e.g., $A_r = \{ T_a | a \in A \}$ where $T_a : R \rightarrow R$ by $xT_a = xa$). Now $R_e \otimes_F A_r$ is a simple Artinian ring and is isomorphic to $R_e A_r$; similarly $R_e \otimes_F B_r$ is isomorphic to $R_e B_r$.

The mapping $R_e \otimes A_r \rightarrow R_e \otimes B_r$ taking $L_r \otimes T_a \rightarrow L_r \otimes T_{\phi(a)}$ is clearly an isomorphism of $R_e \otimes A_r$ onto $R_e \otimes B_r$. We make R into an $R_e A_r$ module by identifying $R_e \otimes A_r$ with $R_e A_r$, that is, $x(L_u \otimes T_a) = uxa$ and into an $R_e \otimes B_r$ module by defining $x(L_u \otimes T_{\phi(a)}) = ux\phi(a)$.

Since $R_e \otimes A_r$ is a simple Artinian ring, by the previous two lemmas R is a direct sum of irreducible $R_e \otimes A_r$-modules V_i which are all isomorphic to each other. Similarly, R is a direct sum of irreducible $R_e B_r$-modules U_i which are isomorphic to each other. Since $R_e \otimes A_r \approx R_e \otimes B_r$ the U_i's are also isomorphic to the V_j's. Now if $R = V_1 \oplus \cdots \oplus V_n$ and $R = U_1 \oplus \cdots \oplus U_m$ with say, $n \leq m$, then we can find module isomorphisms σ_i, $i = 1, 2, \cdots, n$, of V_i onto U_i such that $\sigma_i(v_i L_u T_a) = \sigma_i(v_i) L_u T_{\phi(a)}$ for $v_i \in V_i$, $u \in R$, $a \in A$. Let $\sigma = \sum \sigma_i$; then σ is a monomorphism of R into R with the property that $\sigma(v L_u T_a) = \sigma(v) L_u T_{\phi(a)}$ for all $v \in R$, $u \in R$, $a \in A$.

In particular, for $v = 1$, $a = 1$ we have $\sigma(u) = \sigma(1)L_u = u\sigma(1) = ux$ for all $u \in R$, where $x = \sigma(1)$. However, for $u = 1$, $a \in A$, $v = 1$ we get $\sigma(a) = \sigma(1)T_{\phi(a)} = \sigma(1)\phi(a) = x\phi(a)$. Hence putting $u = a$ in the above calculations yields that $\sigma(a) = ax$; together with $\sigma(a) = x\phi(a)$ we see that $x\phi(a) = ax$. To finish the proof we need but show that x is invertible in R.

Since σ is a monomorphism of R into R, were $vx = 0$ for some $v \neq 0$ in R, then from $0 = vx = v\sigma(1) = \sigma(v)$ we would have $v = 0$. Thus x is not a zero-divisor in R. Since R is Artinian it is a simple exercise to show that an element in R which is not a zero-divisor is invertible in R. Hence x^{-1} exists in R and we have that $\phi(a) = x^{-1}ax$, the desired conclusion.

COROLLARY. *Let A be a simple algebra finite dimensional over its center. Then any automorphism of A leaving the center elementwise fixed is inner.*

Proof. In the theorem put $R = A = B$; if ϕ is the automorphism then $\phi(a) = x^{-1}ax$ for all $a \in A$.

Theorem 4.3.1 is an extremely basic and useful result and we shall have many occasions to use it. We now digress to give it some direct applications to results which have independent interest.

(1) Derivations of simple algebras.

DEFINITION. *An additive map δ of a ring R into itself is called a derivation if $\delta(ab) = a\delta(b) + \delta(a)b$. We say that δ is an inner derivation if there exists a $c \in R$ such that $\delta(x) = xc - cx$ for all $x \in R$.*

For algebras A over a field F we further insist that δ satisfy $\delta(\alpha x) = \alpha\delta(x)$ for $\alpha \in F$, $x \in A$.

PROPOSITION. *Let A be a simple algebra finite dimensional over its center F. Then any derivation of A is inner.*

Proof. Let A_2 be the ring of 2×2 matrices over A; A_2 is simple, has F as center and is finite dimensional over F. Let

$$B = \left\{ \begin{pmatrix} a & \delta(a) \\ 0 & a \end{pmatrix} \middle| a \in A \right\}$$

where δ is a derivation of A. Let

$$C = \left\{ \begin{pmatrix} a & 0 \\ 0 & a \end{pmatrix} \middle| a \in A \right\}.$$

For $\alpha \in F$ it is trivial to show that $\delta(\alpha) = 0$, hence the mapping $\psi: C \to B$ defined by

$$\Psi \begin{pmatrix} a & 0 \\ 0 & a \end{pmatrix} = \begin{pmatrix} a & \delta(a) \\ 0 & a \end{pmatrix}$$

is easily shown to be an isomorphism of C onto B leaving F fixed. Also $C \approx A$. All the conditions of Theorem 4.3.1 are satisfied hence there is an invertible matrix

$$\begin{pmatrix} x & y \\ z & w \end{pmatrix}$$

such that

$$\begin{pmatrix} a & \delta(a) \\ 0 & a \end{pmatrix} \begin{pmatrix} x & y \\ z & w \end{pmatrix} = \begin{pmatrix} x & y \\ z & w \end{pmatrix} \begin{pmatrix} a & 0 \\ 0 & a \end{pmatrix}.$$

Hence:

$$ax + \delta(a)z = xa$$
$$ay + \delta(a)w = ya$$
$$aw = wa$$
$$az = za,$$

for all $a \in A$. These relations imply that w, z are scalars and since

$$\begin{pmatrix} x & y \\ z & w \end{pmatrix}$$

is invertible, one of these, say z, is not 0. Putting $u = xz^{-1}$ we get $\delta(a) = au - ua$ for all $a \in A$, hence δ is inner.

(2) *Wedderburn's Theorem on Finite Division Rings.*

Let D be a finite division ring with center Z and let $[D:Z]=n^2$. If K is a maximal subfield of D then, by Theorem 4.2.2, $[K:Z]=n$. Hence K is a finite field with q^n elements where q is the number of elements in Z. Therefore all the maximal subfields are isomorphic, by an isomorphism leaving Z fixed. By Theorem 4.3.1, if K_1 and K_2 are maximal subfields of D, then there is an $x \in D$ such that $K_2 = x^{-1}K_1x$. If $a \in D$ then $Z(a)$ is a subfield of D hence is imbeddable in a maximal subfield. Thus $D = \bigcup_{x \neq 0 \in D} x^{-1}Kx$, where K is any maximal subfield of D. We leave it to the reader to prove that this is impossible unless $K = D$. (Use that a finite group cannot be the union of the conjugates of a proper subgroup.)

(3) *Frobenius' Theorem on Real Division Algebras.*

Let D be a finite-dimensional, noncommutative division algebra over the real field F. As we noted earlier, $[D:F]=4$.

Let K be a maximal subfield of D; K is isomorphic to the complex numbers, $K = F + Fi$ where $i^2 = -1$. The mapping $\phi: K \to K$ defined by $\phi(\alpha + \beta i) = \alpha - \beta i$ is an automorphism of K onto K leaving F fixed. By the Noether-Skolem theorem $\phi(\alpha + \beta i) = x^{-1}(\alpha + \beta i)x$ for some $x \in D$; hence $\phi(i) = x^{-1}ix = -i$. Thus $x^2i = ix^2$ whence $x^2 \in C(K) = K$; since $x^2 \in \{a \in D \mid ax = xa\}$ and $x \notin F$, we get indeed that $x^2 \in F$, so is real. If $x^2 > 0$ then $x \in F$ follows since all elements positive in F have square roots in F. Hence $x^2 = -\alpha^2$, $\alpha \in F$. Let $j = x/\alpha$; then $j^2 = -1$, $ji = -ij$. The elements $1, i, j, k = ij$ are linearly independent over F so, since $[D:F]=4$, they span D. We have proved a classic result due to Frobenius, namely the

PROPOSITION. *If D is a noncommutative division alge-*

bra finite-dimensional over the real field then D is isomorphic to the division algebra of quaternions.

(4) A theorem of Dickson.

Let D be a division ring with center Z and suppose that $a, b \in D$ are algebraic over Z and satisfy the same minimal polynomial over Z. Since this polynomial is irreducible, $Z(a) \approx Z(b)$ by an isomorphism leaving Z fixed and taking a into b. By the Noether-Skolem Theorem there is an $x \in D$ such that $b = x^{-1}ax$.

Note that this result immediately implies that of Lemma 3.1.1.

(5) A theorem of Albert.

Let D be an ordered division algebra which is algebraic over its center Z. Since $1 > 0$ (for $1 = 1^2$), $n = 1 + 1 + \cdots + 1 > 0$ hence Z is of characteristic 0. We claim that D is commutative, that is, $D = Z$. Suppose not; let $a \in D$, $a \notin Z$. Let $p(x) = x^m + \alpha_1 x^{m-1} + \cdots + \alpha_m$ be the minimal polynomial over Z satisfied by a. Now $b = a - (\alpha_1/m) \notin Z$ satisfies, as minimal polynomial, $q(x) = x^m + \beta_2 x^{m-2} + \cdots + \beta_m$, $\beta_i \in F$. Now it can be proved (and requires a proof) that we can factor $q(x)$ as $q(x) = (x - b)(x - b_2) \cdots (x - b_m)$ for some choice of b_i and that we can permute the factors cyclically.

The b_i as roots of $q(x)$ are conjugates of b; this is the theorem of Dickson just proved. Thus $b_i = d_i^{-1}bd_i$; if $b > 0$ it follows that $b_i > 0$, if $b < 0$ then $b_i < 0$. But $b + b_2 + \cdots + b_n = 0$, a contradiction. From this we get the result of Albert,

PROPOSITION. *An ordered division algebra algebraic over its center must be commutative.*

We now turn to another classic result in the theory of algebras; it is usually called the *double centralizer* or *double commutator* theorem. We recall the notation: if

$S \subset R$ then $C_R(S) = \{x \in R \mid xs = sx \text{ for all } s \in S\}$. Clearly $C_R(S)$ is a subring of R and $S \subset C_R(C_R(S))$. Our theorem gives a condition that $S = C_R(C_R(S))$.

THEOREM 4.3.2. *Let R be a simple Artinian ring with center F and let $A \subset R$ be a finite-dimensional simple sub-algebra over F and containing it. Then:*
1. $C_R(A)$ *is simple,*   2. $C_R(C_R(A)) = A$.

Proof. We consider A acting on A as a ring of linear transformations via the right regular representation. If $n = \dim {}_F A$ then we have imbedded, by this means, A in F_n as A_r. As we have seen several times earlier, $C_{F_n}(A_r) = A_e$ and so $C_{F_n}(C_{F_n}(A_r)) = C_{F_n}(A_e) = A_r$. Since A_e is anti-isomorphic to A it is simple hence $C_{F_n}(A_r)$ is simple. In other words in $R = F_n$ the result holds true.

Now $R \otimes_F F_n$ is a simple Artinian ring and its center is F. Since $A \subset F_n$ (considering A as A_r) and $A \subset R$ then $A \otimes 1$ and $1 \otimes A$ are in $R \otimes F_n$, are finite-dimensional over F, simple and isomorphic by an isomorphism leaving F fixed. By the Noether-Skolem Theorem they are conjugate in $R \otimes F_n$, hence their centralizers in $R \otimes F_n$ are conjugate by the same element. Now

$$C_{R \otimes F_m}(A \otimes 1) = C_R(A) \otimes F_n \quad \text{and}$$
$$C_{R \otimes F_n}(1 \otimes A) = R \otimes C_{F_n}(A);$$

being conjugate in $R \otimes F_n$ their centralizers also are. But

$$C_{R \otimes F_n}(C_{R \otimes F_n}(A \otimes 1))$$
$$= C_{R \otimes F_n}(C_R(A) \otimes F_n) = C_R(C_R(A)) \otimes F$$
$$\approx C_R(C_R(A)) \quad \text{and} \quad C_{R \otimes F_n}(C_{R \otimes F_n}(1 \otimes A))$$
$$= C_{R \otimes F_n}(R \otimes F_n(A))$$
$$= Z(R) \otimes C_{F_n}(C_{F_n}(A)) = F \otimes A \approx A$$

by the argument of the first paragraph. Thus $C_R(C_R(A))$

and A are isomorphic and so are of the same dimension over F. Since $A \subset C_R(C_R(A))$ and they are of the same dimension the equality $A = C_R(C_R(A))$ results. This is part (2) of the theorem.

Now we have seen that $C_R(A) \otimes F_n$ and $R \otimes C_{F_n}(A)$ are conjugate, hence isomorphic. Since $C_{F_n}(A) \approx A^*$ is anti-isomorphic to A, $R \otimes C_{F_n}(A) \approx R \otimes A^*$ is simple. Thus $C_R(A) \otimes F_n$ is simple. In consequence $C_R(A)$ must be simple. This is part (1) of the theorem. The theorem is now completely established.

Note that if R is finite-dimensional over F then $[R:F] = [A:F][C_R(A):F]$ follows from $C_R(A) \otimes F_n = R \otimes A_e$ where $n = [A:F]$. Comparing dimensions yields $[C_R(A):F][A:F]^2 = [R:F][A:F]$ and hence $[R:F] = [A:F][C_R(A):F]$ as stated. In particular, if A is a subfield of R such that $[A:F]^2 = [R:F]$ then $A = C_R(A)$ and so A is a maximal commutative subring of R.

The double centralizer theorem enables us to make a sharper enquiry into the nature of maximal subfields of division algebras. This is the content of

THEOREM 4.3.3. *Let D be a division algebra finite-dimensional over its center F. Then D has a maximal subfield which is separable over F.*

Proof. If $D = F$ there is nothing to prove. If $D \neq F$, invoking the Jacobson-Noether theorem (Theorem 3.2.1) we know that there is an element $a \in D$, $a \notin F$ which is separable over F. We therefore have separable subfields of D which properly contain F. Let K be a subfield of D which contains F, is separable over F and is maximal with this property. We claim that K is a maximal subfield of D; to prove this we must merely show that $K = C_D(K)$.

Now, since K is commutative, $K \subset C_D(K)$; since K is

simple and finite-dimensional over F, applying Theorem
4.3.2 we have that $K = C_D(C_D(K))$. Thus K is the center
of $C_D(K)$. If $C_D(K) \neq K$, by the Jacobson-Noether
Theorem there would be an element $u \in C_D(K)$, $u \neq K$
separable over K. But then $K(u)$ is separable over F and
is strictly larger than K. This contradiction allows us to
conclude that $K = C_D(K)$ and so, that K is a maximal
subfield of D.

We make the

DEFINITION. *If D is a division ring with center F then a
field $K \supset F$ is said to be a splitting field for D if $D \otimes K$ is a
dense ring of linear transformations on a vector space
over K.*

Note that Theorem 4.2.1 assures us that any maximal
subfield of D is a splitting field for D, thus such fields
certainly exist. Note that if $[D:F] < \infty$ then $K \supset F$ is a
splitting field if and only if $D \otimes_F K \approx K_n$. We define a
splitting field for simple algebras in the same terms.

As a simple consequence of Theorem 4.3.3 we have

COROLLARY 1. *If A is a finite-dimensional central
simple algebra over F then A has a separable splitting field.*

Proof. By Wedderburn's theorem $A \approx D \otimes_F F_n$ where
D is a finite-dimensional central division algebra over F.
Let K be a maximal separable subfield of D; then $A \otimes_F K$
$\approx (D \otimes_F F_n) \otimes_F K \approx (D \otimes_F K) \otimes_F F_n \approx F_m \otimes F_n \approx F_{mn}$. K is
the required separable splitting field of A.

COROLLARY 2. *If A is a finite-dimensional central
simple algebra over F then A has a splitting field which is
normal over F.*

Proof. Let K be a separable splitting field for A and
let L be a finite normal extension of F containing K.

Hence $A \otimes_F L \approx (A \otimes_F K) \otimes_K L \approx K_n \otimes L \approx L_n$. Therefore L splits A.

One of the famous open questions in algebra is whether the following sharpened form of Theorem 4.3.3 is true: does any finite-dimensional central division algebra over F contain a maximal subfield which is a *normal* extension of F?

4. Crossed products. Let A be a finite-dimensional central simple algebra over F; by Wedderburn's theorem $A \approx D \otimes F_k$ for D a finite-dimensional central division algebra over F and some integer k. If K is a maximal separable subfield of D then $[K:F] = n$ where $[D:F] = n^2$. Let L be the normal closure of K and suppose that $[L:K] = m$. If $B = D \otimes F_m$ then B is central simple over F and, moreover, $L \subset B$. This is seen merely by noting that $K \otimes F_m = K_m$ is contained in B and L is contained, by means of its regular representation over K, in K_m. Now $[B:F] = [D:F][F_m:F] = n^2m^2 = [L:F]^2$. Therefore in the same class as A in the Brauer group $B(F)$ we have found an algebra B with a maximal subfield L, normal over F, such that $[B:F] = [L:F]^2$. Hence in order to study central simple algebras over F, as far as equivalence in the Brauer group sense is concerned, we may content ourselves with studying central simple algebras A over F such that they have a maximal subfield K normal over F and such that $[A:F] = [K:F]^2$.

Let A be such an algebra and suppose that $[K:F] = n$. Let G be the Galois group of K over F. For $k \in K$ and $\sigma \in G$ we shall write k^σ for $\sigma(k)$, that is, $\sigma(k) = k^\sigma$. By the Noether-Skolem theorem (Theorem 4.3.1) there is an invertible element $x_\sigma \in A$ such that $k^\sigma = x_\sigma^{-1} k x_\sigma$ for every $k \in K$. We leave it to the reader to prove that the x_σ are linearly independent over K in that $\sum_{\sigma \in G} x_\sigma k_\sigma = 0$, $k_\sigma \in K$, forces each $k_\sigma = 0$. However, the linear span over

K of the x_σ's has dimension n^2 over F, hence must be all of A. In short, $A = \left\{ \sum_{\sigma \in G} x_\sigma k_\sigma \mid k_\sigma \in K \right\}$.

Let σ, $\tau \in G$ and $k \in K$; then

$$x_\tau^{-1} x_\sigma^{-1} k x_\sigma x_\tau = x_\tau^{-1} k^\sigma x_\tau = k^{\sigma\tau} = x_{\sigma\tau}^{-1} k x_{\sigma\tau}.$$

This says that $x_{\sigma\tau}(x_\sigma x_\tau)^{-1} \in C_A(K) = K$, in other words, $x_\sigma x_\tau = x_\sigma f(\sigma, \tau)$ where $f(\sigma, \tau) \neq 0$ is in K. Let K' be the set of nonzero elements of K; we study the properties of the mapping $f: G \times G \to K'$.

Since A is an associative algebra, for σ, τ, ν in G we have $x_\sigma(x_\tau x_\nu) = (x_\sigma x_\tau)x_\nu$. Computing this explicitly yields that $f(\sigma, \tau\nu)f(\tau, \nu) = f(\sigma\tau, \nu)f(\sigma, \tau)^\nu$—the so-called *factor set condition*. If $a = x_1 f(1, 1)^{-1}$ then a simple computation reveals that a is a unit element for A, hence $a = 1$. Motivated by this discussion we start all over again introducing functions that behave like f.

DEFINITION. *Let K be a normal extension of F with Galois group G. A function $f: G \times G \to K'$ is called a factor set on G in K if, for all σ, τ, ν in G, $f(\sigma, \tau\nu)f(\tau, \nu) = f(\sigma\tau, \nu)f(\sigma, \tau)^\nu$.*

Since K and G will be fairly fixed in what follows we shall merely refer to f as a factor set. Putting $\tau = \nu = 1$ in the factor set condition shows that $f(\sigma, 1) = f(1, 1)$ for all $\sigma \in G$; putting $\sigma = \tau = 1$ gives us $f(1, \nu) = f(1, 1)^\nu$.

DEFINITION. *Let K be a normal extension of F with Galois group G and let f be a factor set. The algebra $A = (K, G, f)$ is called the crossed product of K and G re f if $A = \left\{ \sum_{\sigma \in G} x_\sigma k_\sigma \mid k_\sigma \in K \right\}$, where equality and addition in A are component-wise and where:*

(1) $k x_\sigma = x_\sigma k^\sigma$ *for* $k \in K$

(2) $x_\sigma x_\tau = x_{\sigma\tau} f(\sigma, \tau)$ *for* σ, $\tau \in G$.

An explicit computation shows that $x_1 f(1, 1)^{-1}$ acts as a unit element for (K, G, f) and that each x_σ is invertible. Also (K, G, f) is associative, its center is F (that is, $F x_1 f(1, 1)^{-1}$ and that it is of dimension n^2 over F where $n = [K : F] = o(G)$.

We also note that (K, G, f) is a simple algebra. Let $U \neq (0)$ be an ideal of (K, G, f) and let $0 \neq u = \sum x_\sigma k_\sigma$ be of shortest length in U. By multiplying by x_σ^{-1} from the left we may suppose that $k_1 \neq 0$. For any $k \in K$, $ku - uk$ is in U; but calculating this out we get $\sum_{\sigma \in G} x_\sigma (k^\sigma - k) k_\sigma \in U$. Since for $\sigma = 1$, $k^\sigma - k = 0$ we have produced an element of shorter length in U; but then $\sum_{\sigma \in G} x_\sigma (k^\sigma - k) k_\sigma = 0$ for all $k \in K$. Since K is normal over F we get from this that $k_\sigma = 0$ for $\sigma \neq 1$. Hence $u = x_1 k_1 \in U$ where $k_1 \neq 0 \in K$; since u is invertible we get that $U = (K, G, f)$ and so (K, G, f) is simple.

We summarize these remarks in

THEOREM 4.4.1. *Let K be a normal extension of F with Galois group G and let f be a factor set. Then the crossed product (K, G, f) is a central simple algebra over F. Moreover, given any central simple algebra A over F then we can find a K, G, f so that, in $B(F)$, $[A] = [(K, G, f)]$.*

Note that the open question about the existence of normal maximal subfields in central division algebras becomes, in these terms, if D is a central division algebra over F is then $D = (K, G, f)$ for some K, G, f (not merely $[D] = [(K, G, f)]$)?

Given K normal over F with Galois group G a natural question immediately presents itself: if f, g are two factor sets when are (K, G, f) and (K, G, g) isomorphic? If the $x_\sigma \in A = (K, G, f)$ multiply via $x_\sigma x_\tau = x_{\sigma\tau} f(\sigma, \tau)$ and $x_\sigma^{-1} k x_\sigma = k^\sigma$ then for any choice $\lambda_\sigma \in K'$ the elements $y_\sigma = x_\sigma \lambda_\sigma$ span A, induce the automorphisms of G on K and multiply according as $y_\sigma y_\tau = x_\sigma \lambda_\sigma x_\tau \lambda_\tau = x_{\sigma\tau} \lambda_\sigma^\tau \lambda_\tau$

$= x_{\sigma\tau}f(\sigma,\ \tau)\lambda_\sigma{}^\tau\lambda_\tau = y_{\sigma\tau}\lambda_{\sigma\tau}{}^{-1}\lambda_\sigma{}^\tau\lambda_\tau f(\sigma,\ \tau)$. This shows that $\lambda_{\sigma\tau}{}^{-1}\lambda_\sigma{}^\tau\lambda_\tau f(\sigma,\tau)$ is a factor set and gives rise to an algebra isomorphic to A. If Ψ is an automorphism of $B = (K,\ G,\ g)$ onto $A = (K,\ G,\ f)$ by following Ψ by an inner automorphism of A and an automorphism of K we may assume that $\Psi(k) = k$ for all $k \in K$. If the $z_\sigma \in B$ induce the σ on K and $z_\sigma z_\tau = z_{\sigma\tau}g(\sigma,\ \tau)$ then $\Psi(z_\sigma) = y_\sigma$ induces σ on K in A so that $\Psi(z_\sigma) = y_\sigma = x_\sigma\lambda_\sigma$ for $\lambda_\sigma \in K'$. This yields that $g(\sigma,\ \tau) = \lambda_{\sigma\tau}{}^{-1}\lambda_\sigma\lambda_\tau f(\sigma,\ \tau)$. Conversely, given the factor sets f, g for which we can find a function $\lambda : G \to K'$ such that $g(\sigma,\ \tau) = (\lambda_\sigma{}^\tau\lambda_\tau/\lambda_{\sigma\tau})f(\sigma,\ \tau)$ then reversing the above discussion we get that $(K,\ G,\ f)$ and $(K,\ G,\ g)$ are isomorphic as algebras over F. This motivates the

DEFINITION. *Two factor sets f, g are equivalent if there exists a function $\lambda : G \to K'$ such that $g(\sigma,\ \tau) = (\lambda_\sigma{}^\tau\lambda_\tau/\lambda_{\sigma\tau})$ $\cdot f(\sigma,\tau)$ for all $\sigma, \tau \in G$.*

In the argument prior to the definition we proved

LEMMA 4.4.1. *The algebras $(K,\ G,\ f)$ and $(K,\ G,\ g)$ are isomorphic if and only if f and g are equivalent.*

In view of the lemma the equivalence of factor sets is indeed an equivalence relation. Given a factor set f let $t = f(1,\ 1)^{-1}$ and let $\lambda_\sigma = t^\sigma$; then $g(\sigma,\ \tau) = (\lambda_\sigma{}^\tau\lambda_\tau/\lambda_{\sigma\tau})$ $f(\sigma,\ \tau) = t^\tau f(\sigma,\ \tau)$ satisfies

$$g(\sigma,\ 1) = g(1,\ 1) = t^{-1}f(1,\ 1) = 1$$

and

$$g(1,\ \sigma) = t^\sigma f(1,\ \sigma) = \left(f(1,\ 1)^{-1}\right)^\sigma f(1,\ 1)^\sigma = 1.$$

Since g is equivalent to f it gives an isomorphic algebra. Thus in studying crossed products we may, without loss of generality, assume that our factor sets are normalized in the sense that $f(\sigma,\ 1) = f(1,\ \sigma) = 1$ for all $\sigma \in G$.

Let f, g be factor sets; we define $h(\sigma, \tau) = f(\sigma, \tau)g(\sigma, \tau)$ for all σ, $\tau \in G$. Calculating we get that h is a factor set; we call it the *product* of f and g and write it as $h = fg$. We should like to interrelate the three algebras (K, G, f), (K, G, g) and (K, G, fg).

Under the product defined the factor sets form a group; the unit element in this group is the factor set f defined by $f(\sigma, \tau) = 1$ for all σ, $\tau \in G$ and the inverse of the factor set g is defined, most naturally, by $g^{-1}(\sigma, \tau) = (g(\sigma, \tau))^{-1}$. The factor sets g in this group equivalent to 1 form a subgroup, this follows easily from the fact that $g(\sigma, \tau) = (\lambda_\sigma^\tau \lambda_\tau)/\lambda_{\sigma\tau}$. The quotient group of the group of factor sets modulo the ones equivalent to 1 is a familiar object in homological algebra being $H^2(G, K')$; we shall discuss this relation a little more later. Note that this group is in one-to-one correspondence with the isomorphism classes of (K, G, f).

Our objective is to prove the following fundamental result which intertwines the product in the Brauer group and the product of factor sets: $[(K, G, f)]$ $[(K, G, g)] = [(K, G, fg)]$. But first we develop some subsidiary results which themselves are interesting theorems.

LEMMA 4.4.2. $(K, G, 1) \approx F_n$.

Proof. Let $[K : F] = n$, hence $[(K, G, 1) : F] = n^2$. Now $(K, G, 1)$ has K as a subfield and, for each $\sigma \in G$, elements x_σ such that $x_\sigma^{-1} k x_\sigma = k^\sigma$ and $x_\sigma x_\tau = x_{\sigma\tau}$. We let $(K, G, 1)$ act on K as follows: for $x = \sum x_\sigma k_\sigma$ in $(K, G, 1)$ and $k \in K$ we define $kR_x = \sum k^\sigma k_\sigma$. This defines a linear transformation on K over F. Our defining relations for $(K, G, 1)$ assure us that the mapping $(K, G, 1) \to A_F(K)$, the ring of F-linear transformations on K, defined by $x \to R_x$ is a nonzero homomorphism. Since $(K, G, 1)$ is simple this mapping must be a monomorphism. However, since $(K, G, 1)$ is of dimension n^2 over F, as is $A_F(K)$ we get that this monomorphism is onto hence is an isomorphism. Since $A_F(K) \approx F_n$ we have proved that $(K, G, 1) \approx F_n$.

The next result points out a very interesting feature of central simple algebras, namely, whenever they sit in a larger algebra they do so in a very particular way. In fact the property of the theorem characterizes finite-dimensional central simple algebras. However we only prove the result in one direction.

THEOREM 4.4.2. *Let B be an algebra with unit element over a field F and suppose that $B \supset A$ where A is a finite-dimensional central simple algebra over F having the same unit element as B. Then $B = A \otimes_F C_B(A)$.*

Proof. We first settle the special case $A = F_m$. Let e_{ij} be the usual matrix units in F_m. If $b \in B$ let $b_{ij} = \sum_{k=1}^m e_{ki} b \, e_{jk}$; making the calculation we verify that $b_{ij} \in C_B(F_m)$. Now

$$\sum_{i,j} b_{ij} e_{ij} = \sum_{i,j,k} e_{ki} b e_{jk} e_{ij} = \sum_{i,j} e_{ii} b e_{jj}$$

$$= \left(\sum_i e_{ii} \right) b \left(\sum_j e_{jj} \right) = b$$

since $\sum_i e_{ii} = 1$, hence we have that $B \subset C_B(F_m) F_m \subseteq B$ and so $B = F_m C_B(F_m)$. To show that $B = F_m \otimes_F C_B(F_m)$ we must merely prove that if $b = \sum b_{ij} e_{ij} = 0$ with $b_{ij} \in C_B(F_m)$ then $b_{ij} = 0$ for all i, j. But, as above, we get $b_{ij} = \sum_k e_{ki} b e_{jk} = 0$. Therefore $B = F_m \otimes_F C_B(F_m)$; note that this says that $B = C_m$ where $C = C_B(F_m)$.

Now let A be any finite-dimension central simple algebra, $B \supset A$. Hence $E = B \otimes_F A^* \supset A \otimes_F A^* = F_m$ (by Theorem 4.1.3) and E and F_m have the same unit element. By the argument of the first paragraph we have

$$E = B \otimes {}_F A^* = C_E(F_m) \otimes {}_F F_m = (C_E(F_m) \otimes {}_F A) \otimes {}_F A^*.$$

Because A^* is central simple and $B = C_E(A^*)$ from $E = B \otimes {}_F A^*$ whereas $C_E(F_m) \otimes {}_F A = C_E(A^*)$ from $E = (C_E(F_m) \otimes {}_F A) \otimes {}_F A^*$ we get that $B = C_E(F_m) \otimes {}_F A$. From this representation it is clear that $C_E(F_m) = C_B(A)$ and so the theorem is proved.

To reach our immediate goal we need another lemma; its result, also, is of great independent interest.

LEMMA 4.4.3. *Let K be a finite normal extension of F with Galois group G. Then $K \otimes {}_F K = \oplus_{\sigma \in G} e_\sigma (K \otimes 1)$ $= \oplus_{\sigma \in G} e_\sigma (1 \otimes K)$ where the e_σ are orthogonal idempotents such that, for all $k \in K$, $e_\sigma(k \otimes 1) = e_\sigma(1 \otimes k^\sigma)$.*

Proof. The proof can be given in many forms; the one given here is chosen because of its explicit nature.

If K is merely separable then $K = F(a)$ for some a where a is a root of its minimal polynomial $p(x)$ which is not divisible by any square of a polynomial, hence

$$K \otimes {}_F K \approx K \otimes \frac{F[x]}{(p(x))} \approx \frac{K[x]}{(p(x))},$$

so is semisimple since it can have no nilpotent elements.

If K is normal let $K = F(a)$, $[K : F] = n = o(G)$, the Galois group of K and F. Let the minimal polynomial for a over F be such that $a^n + \alpha_1 a^{n-1} + \cdots + \alpha_n = 0$. If $\sigma \in G$ let $b_{\sigma,m} = a^m \otimes 1 + a^{m-1} \otimes a^\sigma + \cdots + 1 \otimes (a^\sigma)^m$; from the fact that $1, a \cdots, a^{n-1}$ (and $1, a^\sigma, \cdots, (a^\sigma)^{n-1}$) are linearly independent over F we get that $b_{\sigma,1}, \cdots, b_{\sigma,n-1}$ are linearly independent over F in $K \otimes K$. Moreover

(1) $(a \otimes 1 - 1 \otimes a^\sigma) b_{\sigma,m} = a^{m+1} \otimes 1 - 1 \otimes (a^\sigma)^{(m+1)}.$

From (1) and the polynomial for a we get

(2) $(a \otimes 1 - 1 \otimes a^\sigma)(b_{\mu,n-1} + \alpha_1 b_{\sigma,n-2} + \cdots$
$$+ \alpha_{n-2} b_{\tau,1}) = 0$$

hence $a \otimes 1 - 1 \otimes a^\sigma$ is a zero divisor in $K \otimes K$. Thus there is a minimal idempotent e_σ such that $e_\sigma(a \otimes 1 - 1 \otimes a^\sigma)$ $= 0$. If $x \in K$ then $x = \sum_{i=0}^{n-1} \beta_i a^i$, $\beta_i \in F$ hence

$$e_\sigma(x \otimes 1 - 1 \otimes x^\sigma) = e_\sigma \left(\sum_{i=0}^{n-1} \beta_i(a^i \otimes 1 - 1 \otimes (a^\sigma)^i) \right)$$

$$= e_\sigma(a \otimes 1 - \otimes a^\sigma) \sum \beta_i b_\sigma = 0.$$

Thus indeed $e_\sigma(k \otimes 1) = e_\sigma(1 \otimes k^\sigma)$ for $k \in K$. Now by comparing dimensions and from the obvious fact that $e_\sigma \neq e_\tau$ for $\sigma \neq \tau$ in G we get, using the mimimality of e_σ,

$$\sum_{\sigma \in G} e_\sigma(K \otimes K) = \sum_{\sigma \in G} e_\sigma(K \otimes 1) = K \otimes K.$$

Since distinct minimal idempotents annihilate each other $e_\sigma e_\tau = 0$ for $\sigma \neq \tau$ in G.

We are now ready to establish the relation between the product in $B(F)$ and that of factor sets. But first we need the

SUBLEMMA. *If A is central simple over F and $e \neq 0$ is an idempotent in A then in $B(F)$, $[A] = [eAe]$ (here we are identifying F and Fe).*

Proof. By Wedderburn's theorem $A = D_m$ and $[A]$ $= [D]$; by a change of basis, that is, an inner automorphism of A, we may assume that

$$e = \begin{pmatrix} I_r & 0 \\ 0 & 0 \end{pmatrix}$$

where I_r is the $r \times r$ identity matrix. Thus

$$eAe = \begin{pmatrix} I_r & 0 \\ 0 & 0 \end{pmatrix} D_m \begin{pmatrix} I_r & 0 \\ 0 & 0 \end{pmatrix} = \begin{pmatrix} D_r & 0 \\ 0 & 0 \end{pmatrix},$$

hence $eAe \approx D_r$ and so $[eAe] = [D] = [A]$.

Now to

THEOREM 4.4.3. *If K is normal over F with Galois group G and if f, g are factor sets then $[(K, G, f)][(K, G, g)]$ $= [(K, G, fg)]$. Equivalently, $(K, G, f) \otimes_F (K, G, g)$ $\approx (K, G, fg) \otimes_F F_m$.*

Proof. Let $E = (K, G, f) \otimes_F (K, G, g)$. Since $E \supset K \otimes K$, applying Lemma 4.4.3 we get, for each $\sigma \in G$, an idempotent $e_\sigma \neq 0$ in $K \otimes K$ such that $e_\sigma e_\tau = 0$ for $\sigma \neq \tau$ satisfying $e_\sigma(k \otimes 1) = e_\sigma(1 \otimes k^\sigma)$ for $k \in K$. Let $e = e_1$, so that $e(k \otimes 1) = e(1 \otimes k)$.

Let $x_\sigma \in (K, G, f)$ such that $x_\sigma^{-1} k x_\sigma = k^\sigma$ and $x_\sigma x_\tau = x_{\sigma\tau} f(\sigma, \tau)$ and let $y_\sigma \in (K, G, g)$ such that $y_\sigma^{-1} k y_\sigma = k^\sigma$ and $y_\sigma y_\tau = y_{\sigma\tau} g(\sigma, \tau)$.

Consider $(1 \otimes y_\sigma) e (1 \otimes y_\sigma^{-1})$; it is easy to see that this is an idempotent in $K \otimes K$ and, moreover,

$$(1 \otimes y_\sigma) e (1 \otimes y_\sigma^{-1})(1 \otimes k^\sigma) = (1 \otimes y_\sigma) e (1 \otimes k)(1 \otimes y_\sigma^{-1})$$
$$= (1 \otimes y_\sigma) e (k \otimes 1)(1 \otimes y_\sigma^{-1})$$
$$= (1 \otimes y_\sigma) e (1 \otimes y_\sigma^{-1})(k \otimes 1).$$

In view of this we have $(1 \otimes y_\sigma) e (1 \otimes y_\sigma^{-1}) = e_\sigma$; in particular, for $\sigma \neq 1$, $e(1 \otimes y_\sigma) e = e e_\sigma (1 \otimes y_\sigma) = 0$. Similarly $(x_\sigma^{-1} \otimes 1) e (x_\sigma \otimes 1) = e_\sigma$. Thus if $w_\sigma = x_\sigma \otimes y_\sigma$ then $w_\sigma e = e w_\sigma = e w_\sigma e$.

Let $u_\sigma = e w_\sigma$; $u_\sigma \in eEe$ and is invertible in eEe having $e w_\sigma^{-1}$ as inverse. Now

$$u_\sigma u_\tau = e w_\sigma e w_\tau = e w_\sigma w_\tau = e(x_\sigma x_\tau \otimes y_\sigma y_\tau)$$
$$= e(x_{\sigma\tau} f(\sigma, \tau) \otimes y_{\sigma\tau} g(\sigma, \tau)$$
$$= e(x_{\sigma\tau} \otimes y_{\sigma\tau}) e(f(\sigma, \tau) \otimes g(\sigma, \tau))$$
$$= u_{\sigma\tau} e(f(\sigma, \tau) g(\sigma, \tau) \otimes 1).$$

Also, since $K \otimes K$ is commutative, $e(K \otimes 1) = (K \otimes 1)e$ and

$$u_\sigma^{-1} e(k \otimes 1) u_\sigma = e(x_\sigma^{-1} \otimes y_\sigma^{-1})(k \otimes 1)(x_\sigma \otimes y_\sigma)$$
$$= e(k^\sigma \otimes 1).$$

Thus $eEe \supset (e(K \otimes 1), G, e(f \otimes 1)(g \otimes 1))$.

Now for k, $k' \in K$ we have

$$e(x_\sigma k \otimes y_\tau k')e = e(x_\sigma \otimes y_\sigma)(1 \otimes y_\sigma^{-1} y_\tau)(k \otimes k')e$$
$$= (x_\sigma \otimes y_\sigma)e(1 \otimes y_\sigma^{-1} y_\tau)e(k \otimes k')$$
$$= u_\sigma e(1 \otimes y_\sigma^{-1} y_\tau)e(kk' \otimes 1).$$

If $\sigma \neq \tau$ then $e(1 \otimes y_\sigma^{-1} y_\tau)e = 0$; if $\sigma = \tau$ then $e(x_\sigma k \otimes y_\sigma k')e = u_\sigma e(kk' \otimes 1)$. In short, we get $eEe \subset \sum u_\sigma e(K \otimes 1) \subset eEe$ and so $eEe = \sum u_\sigma e(K \otimes 1) = (e(K \otimes 1), G, e(fg \otimes 1))$, that is, $eEe \approx (K, G, fg)$.

Now, by the sublemma, $[E] = (eEe)$ hence $[E] = [eEe] = [(K, G, fg)]$ as claimed by the theorem.

Theorem 4.4.3 demonstrates that $\{[(K, G, f)]\}$ as f varies over the vector sets on G in K' is a subgroup of $B(F)$. We can view this in another light. Let K be a normal extension of F; we map $B(F)$ into $B(K)$ by mapping $[A]$ in $B(F)$ into $[A \otimes_F K]$ in $B(K)$. The kernel of this homomorphism is the set of all central simple algebras over F split by K. As we have seen these algebras are all of the form (K, G, f).

To study $B(F)$ we have seen that it suffices to consider $\{[(K, G, f)]\}$ where K runs over all the normal extensions of F, G is the Galois group of K over F and f is a factor set on G to K'. We consider another approach to all this.

Let G be a group and M a right G-module. Let $C^n(G, M)$ be the set of all functions $f : G \times G \times \cdots \times G$

$\rightarrow M$ (the product taken n times). We define a mapping $\delta^n : C^n \rightarrow C^{n+1}$ by means of

$$(\delta^n f)(x_1, \cdots, x_{n+1}) = f(x_2, \cdots, x_{n+1})$$

$$+ \sum_{i=1}^{n} (-1)^i f(x_1, \cdots, x_i x_{i+1}, \cdots, x_{n+1})$$

$$+ (-1)^{n+1} f(x_1, \cdots, x_n) x_{n+1}.$$

One can verify that $\delta^n \delta^{n+1} = 0$. If $Z^n = \{f \in C^n \mid \delta^n f = 0\}$ and $B^n = \{\delta^{n-1} f \mid f \in C^{n-1}\}$ we have that $B^n \subset Z^n$. Let $H^n(G, M) = Z^n / B^n$; we call this the nth *cohomology group* of G in M.

Note that if K is a normal extension of F with Galois group G then K' is a G-module in the natural way (if we write the product in K' as $+$). Note also that a function in two variables from G to K' is a factor set if and only if it is an element of $Z^2(G, K')$ and that two factor sets f, g are equivalent if and only if $fg^{-1} \in B^2(G, K')$. Hence the set of factor sets on G to K' modulo these equivalent to 1 is indeed the group $H^2(G, K')$. Thus $\{[(K, G, f)]\}$ is isomorphic to $H^2(G, K')$.

We prove a result for $H^k(G, M)$ which, in the special case where G is a finite group, has a very important implication for $B(F)$. This is

LEMMA 4.4.4. *If G is a finite group of order $o(G)$ then, for $n > 0$, $o(G) H^n(G, M) = (0)$.*

Proof. Let $f \in Z^n(G, M)$; hence for $x_1, \cdots, x_{n+1} \in G$, $(\delta^n f)(x_1, \cdots, x_{n+1}) = 0$. Writing this out explicitly we have:

$$f(x_2, \cdots, x_{n+1})$$

$$= -\left(\sum_{i=1}^{n} (-1)^i f(x_1, \cdots, x_i x_{i+1}, \cdots, x_{n+1}) \right.$$

$$\left. + (-1)^{n+1} f(x_1, \cdots, x_n) x_{n+1} \right).$$

We sum this over G, letting x_1 take on all values in G. We get

$$o(G)f(x_2, \cdots, x_{n+1})$$

$$= -\left(\sum_{i=1}^{n} (-1)^i \sum_{x_1 \in G} f(x_1, \cdots, x_i x_{i+1}, \cdots, x_{n+1}) \right.$$

$$\left. + (-1)^{n+1} \sum_{x_1 \in G} f(x_1, \cdots, x_n) x_{n+1} \right) \cdot$$

Let $h(x_2, \cdots, x_n) = \sum_{x_1 \in G} f(x_1, x_2, \cdots, x_n)$. Now

$$\sum_{x_1 \in G} f(x_1 x_2, x_3, \cdots, x_{n+1})$$

$$= \sum_{x_1 \in G} f(x_1, x_3, \cdots, x_{n+1}) = h(x_3, \cdots, x_{n+1})$$

since $x_1 x_2$ runs over G as x_1 does. In terms of h the above relation becomes

$$o(G)f(x_2, \cdots, x_{n+1}) = - h(x_3, \cdots, x_{n+1})$$

$$+ \sum_{i=2}^{n+1} (-1)^i h(x_2, \cdots, x_i x_{i+1}, \cdots, x_{n+1})$$

$$- (-1)^{n+1} h(x_2, \cdots, x_n) x_{n+1}$$

$$= -(\delta^{n-1}h)(x_2, \cdots, x_{n+1}) \in B^n(G, \dot{M}).$$

Therefore $o(G)Z^n(G, M) \subset B^n(G, M)$ hence $o(G)H^n(G, M) = (0)$.

This lemma has as an immediate consequence the important

THEOREM 4.4.4. *Every element in $B(F)$ has finite order, that is, $B(F)$ is a torsion group.*

Proof. If $[A] \in B(F)$ then, using Theorem 4.4.1, $[A] = [(K, G, f)]$ for some finite normal extension K of F. Since $[(K, G, f)] \in H^2(G, K')$ we get, from the previous lemma, that $[A]^{o(G)} = [(K, G, f)]^{o(G)} = 1$.

Since every element in $B(F)$ has finite order it can be written as the product of elements of prime power orders. What does this mean about the decomposition of the algebras themselves? To get such a Sylow-type decomposition we need more precise information about the order of an element in $B(F)$ than is given us in Theorem 4.4.4.

If D is a central division algebra over F then $[D:F] = n^2$; we call n the *degree* of D and write it as $\delta_F(D)$. If A is central simple over F then $A \approx D \otimes_F F_n$ where D is a central division algebra over F. We define $\delta_F(A)$ by $\delta_F(A) = \delta_F(D)$. Finally we define $e_F(A)$ to be the order of $[A]$ in $B(F)$. We prove

THEOREM 4.4.5. *If A is central simple over F then* $[A]\delta_F(A) = 1$ *in* $B(F)$, *that is*

$$\underbrace{A \otimes A \otimes \cdots \otimes A}_{\delta_F(A)\text{-times}} = F_m.$$

Equivalently, $e_F(A) \mid \delta_F(A)$.

Proof. In $B(F)$ $[A] = [D]$ for some central division algebra D. If K_0 is a separable maximal subfield of D then $[D:K_0]^2 = n^2 = \delta_F(A)^2 = [D:F]$. Let K be the normal extension of F generated by K_0 and suppose that $[K:K_0] = q$. As we saw earlier $D \otimes F_q = D_q$ contains K and $D_q = (K, G, f)$ where G is the Galois group of K over F and f is a factor set on G to K'.

Now $D_q = \rho_1 \oplus \cdots \oplus \rho_q$ where the ρ_i are minimal right ideals and so are q-dimensional as right vector spaces over D. The ρ_i are right vector spaces over K. Since $[\rho_i:K]q = [D_q:K] = [K:F] = qn$ we get $[\rho_i:K] = n = \delta_F(A)$.

Let $x_\sigma \in D_q = (K, G, f)$ be such that $x_\sigma^{-1} k x_\sigma = k^\sigma$ for $k \in K$ and $x_\sigma x_\tau = x_{\sigma\tau} f(\sigma, \tau)$. Since ρ_1 is a right ideal,

$\rho_1 x_\sigma \subset \rho_1$, hence x_σ induces an endomorphism T_σ on ρ_1. If $u_1, \cdots, u_n$ is a basis of ρ_1 over K then $u_i T_\sigma = \sum_j u_j t_{ij\sigma}$ (where $t_{ij\sigma} \in K$) where $T_\sigma = (t_{ij\sigma})$ is an $n \times n$ matrix over K. Now

$$u_i T_\sigma T_\tau = (\sum u_j t_{ij\sigma}) T_\tau = \sum u_j T_\tau t_{ij\sigma}^\tau;$$

since $T_\sigma T_\tau = T_{\sigma\tau} f(\sigma, \tau)$ we get that $T_\sigma f(\sigma, \tau) = T_\tau T_\sigma^\tau$. Let $\lambda_\sigma = \det T_\sigma$; we have $f(\sigma, \tau)^n \lambda_{\sigma\tau} = \lambda_\tau \lambda_\sigma^\tau$, hence $f(\sigma, \tau)^n$ is equivalent to the unit factor set. Therefore $[(K, G, f)]^n = [F]$. We have proved that $[A]^{\delta_F(A)} = 1$ in $B(F)$.

We dig deeper into the relationship of $e_F(A)$ and $\delta_F(A)$. To do so we need a preliminary remark, namely the

SUBLEMMA. *If D is a finite dimensional algebra over its center F and if K, an extension of F, splits D then $\delta_F(D) \mid [K : F]$.*

Proof. Suppose that $D \otimes_F K = K_r$; computing dimensions over F we get $\delta_F(D)^2 m = mr^2$ where $m = [K : F]$, hence $r = \delta_F(D)$.

Using the regular representation we may suppose that $K \subset F_m$ hence $D \otimes_F F_m \supset D \otimes_F K = K_r \supset F_r$. By Theorem 4.4.2, $D \otimes_F F_m = F_r \otimes_F Y$. From the uniqueness part of Wedderburn's theorem we easily get from this that $r \mid m$, that is, $\delta_F(D) \mid [K : F]$.

LEMMA 4.4.5. *If p is a prime number and $p \mid \delta_F(A)$ then $p \mid e_F(A)$.*

Proof. If L is an extension field of F the mapping $A \to A \otimes_F L$ induces a homomorphism of $B(F)$ into $B(L)$. In consequence $e_L(A \otimes L) \mid e_F(A)$. We can find a normal extension K of F with Galois group G such that $[A] = [(K, G, f)]$ for some factor set f and where $(K, G, f) = D_m$, D a central division algebra F with $\delta_F(D) = \delta_F(A)$ and $e_F(D) = e_F(A)$; $[K : F] = \delta_F(D)m = n$. Since $p \mid \delta_F(A)$

$= \delta_F(D)$ we have that $p \mid n = o(G)$; let G_p be a p-Sylow sub-group of G of order $p^s \neq 1$. By elementary Galois theory K has a subfield K_0 with $[K : K_0] = p^s$. Now since $p \nmid [K_0 : F]$ by the sublemma it is impossible that K_0 split A, yet by assumption p divides the degree of a maximal subfield of D. Therefore $e_{K_0}(A \otimes_F K_0) \neq 1$. Since K splits $A \otimes_F K_0$ and $[K : K_0] = p^s$ by Theorem 4.4.5, $e_{K_0}(A \otimes_F K_0) = p^t \neq 1$. Hence $p \mid e_{K_0}(A \otimes_F K_0) \mid e_F(A)$. This is exactly what is claimed in the lemma.

COROLLARY. *If* $e_F(A) = p^t$, p *a prime number, then* $\delta_F(A) = p^s$ *with* $s \geq t$.

It is not always true that $\delta_F(A) = e_F(A)$. However for certain fields F this is true, for instance if F is an algebraic number field or if F is a p-adic field. This is not an easy result and is important in discussing central division algebras over algebraic number fields. These division algebras have a beautiful description due to a theorem of Albert, Brauer, Hasse and Noether, namely: If D is a central division algebra over an algebraic number field then $D = (K, G, f)$ where G is a cyclic group (and f is determined by a nonzero element in F).

We need one more preliminary result,

LEMMA 4.4.6. *Let* D *and* Δ *be central division algebras over* F. *If* $\delta_F(D)$ *is relatively prime to* $\delta_F(\Delta)$ *then* $D \otimes_F \Delta$ *is a division algebra.*

Proof. Since $D \otimes_F \Delta$ is a central simple algebra over F, by Wedderburn's theorem $D \otimes_F \Delta = E \otimes_F F_m$ where E is a central division algebra over F. We should like to prove that $m = 1$, so that $D \otimes_F \Delta = E$ would result.

Now $D^* \otimes_F D \approx F_n$ where $n = \delta_F(D)^2$. Also $D^* \otimes_F E \approx D_1 \otimes F_r$ where D_1 is a central division algebra; this comes from Wedderburn's theorem and the fact that $D^* \otimes_F E$ is central simple over F. Now

$$F_n \otimes \Delta \approx (D^* \otimes D) \otimes \Delta \approx D^* \otimes (D \otimes \Delta)$$

$$\approx D^* \otimes (E \otimes F_m) \approx D_1 \otimes F_r \otimes F_m$$

$$\approx D_1 \otimes F_{rm}.$$

By the uniqueness part of Wedderburn's theorem $rm = u = \delta_F(D)^2$, hence $m \mid \delta_F(D)^2$. Similarly $m \mid \delta_F(\Delta)^2$. The relative primeness of $\delta_F(D)$ and $\delta_F(\Delta)$ implies that $m = 1$. This proves that $D \otimes \Delta \approx E$ is a division algebra.

We conclude this chapter by giving a Sylow-like decomposition of central division algebras. This is

THEOREM 4.4.6. *If D is a finite-dimensional central division algebra over F with $\delta_F(D) = p_1^{m_1} \cdots p_k^{m_k}$, where the p_i are distinct primes, then $D = D_1 \otimes_F D_2 \cdots \otimes_F D_k$ where $\delta_F(D_i) = p_i^{m_i}$.*

Proof. By Theorem 4.4.5 we have, in $B(F)$, that $[D]^{\delta_F(D)} = 1$ hence we can decompose $[D]$ as $[D] = [A_1] \cdots [A_k]$ where each A_i has order dividing $p_i^{m_i}$. Now $[A_i] = [D_i]$ where D_i is a central division algebra over F; therefore $[D] = [D_1] \cdots [D_i]$. Hence

$$D \otimes F_n \approx (D_1 \otimes F_{n_1}) \otimes (D_2 \otimes F_{n_2}) \otimes \cdots \otimes (D_k \otimes F_{n_k}),$$

which is to say,

$$D \otimes F_n \approx (D_1 \otimes \cdots \otimes D_k) \otimes F_{n_1 n_2 \cdots n_k}.$$

Since $e_F(D_i) = p^{\alpha}$, by Lemma 4.4.5 $\delta_F(D_i) = p_i^{f_i}$. By Lemma 4.4.6, $D_1 \otimes \cdots \otimes D_k$ is a division algebra. Applying the uniqueness part of Wedderburn's theorem we get $D \approx D_1 \otimes \cdots \otimes D_k$. Counting dimensions we obtain $f_i = m_i$, hence $\delta_F(D_i) = p^{m_i}$. This establishes the theorem.

The theorem allows us, in most studies about finite-dimensional division algebras, to reduce to the special case of a division algebra whose division over its center is a power of a prime number.

References

1. A. A. Albert, *Structure of algebras*, Amer. Math. Soc. Colloq. Publ., XXIV (1939).

2. ———, On ordered algebras, *Bull. Amer. Math. Soc.*, 45 (1940) 521–522.

3. A. A. Albert and H. Hasse, A determination of all normal division algebras over an algebraic number field, *Trans. Amer. Math. Soc.*, 34 (1932) 722–726.

4. S. A. Amitsur, Simple algebras and cohomology groups of arbitrary fields, *Trans. Amer. Math. Soc.*, 96 (1959) 73–112.

5. E. Artin, C. Nesbitt and R. Thrall, *Rings with minimum conditions*, Univ. of Michigan, 1944.

6. M. Auslander and O. Goldman, The Brauer group of a commutative ring, *Trans. Amer. Math. Soc.*, 97 (1960) 367–409.

7. R. Brauer, Über Systeme hypercomplexer Zahlen, *Math. Z.* 29 (1929) 79–107.

8. R. Brauer, H. Hasse and E. Noether, Beweis eines Hauptsatzes in der Theorie der Algebren, *Jour. fur Math.*, 107 (1931) 399–404.

9. S. Chase, A. Rosenberg and D. Harrison, *Galois theory and cohomology of commutatative rings*, Amer. Math. Soc. Memoirs, 52 (1965).

10. M. Deuring, *Algebren*, Ergeb. Math., 4 (1935), Springer, Berlin.

11. N. Jacobson, *Theory of rings*, Amer. Math. Soc., Math. Surveys, II (1943) Providence.

12. ———, *Structure of rings*, Amer. Math. Soc. Colloq. Publ., 37 (1964).

13. A. Rosenberg and D. Zelinsky, On Amitsur's complex, *Trans. Amer. Math. Soc.*, 97 (1960) 327–356.

CHAPTER 5

REPRESENTATIONS OF FINITE GROUPS

In Maschke's theorem (Theorem 1.4.1) we showed that the group algebra $F(G)$ of a finite group G of order $o(G)$ over a field F of characteristic 0 or p where $p \nmid o(G)$ is semi-simple. By the theorems we have already proved about the nature of semi-simple Artinian rings the structure of $F(G)$ is fairly decisively pinned down. The information we garner this way about $F(G)$ allows us to probe more deeply in G itself. It is this interplay between G and $F(G)$ and its consequences that we propose to study in this chapter. *We shall assume throughout—unless otherwise stated—that F is the field of complex numbers.* Most of what we do could be done for any algebraically closed field of characteristic 0 or p where $p \nmid o(G)$.

1. The elements of the theory. Cayley's theorem in the theory of finite groups asserts that every finite group is isomorphic to a group of permutations; these permutations in turn have a very nice representation as matrices whose entries are 0's and 1's. Nice as this realization of the group as a group of matrices is there are many nicer and more important ways of representing the group—homomorphically now instead of isomorphically —as a group of matrices.

We begin with the

DEFINITION. *A representation of G is a homomorphism ψ of G into $L(V)$, the algebra of linear transformations on V over F, such that $\psi(1) = I$, the identity transformation.*

We shall call V the *representation module* for G belonging to Ψ. We use the term module advisedly for V carries the structure of a G-module, and hence an $F(G)$-module, by defining the module action via $v \cdot g = v\psi(g)$ for $g \in G$, $v \in V$. Conversely, given an $F(G)$-module M it affords us a representation of G and of $F(G)$; this is done by defining $\psi: F(G) \to L(M)$ via $mg = m\psi(g)$ for $g \in G$, $m \in M$. Hence the study of the representations of G is equivalent to the study of $F(G)$-modules.

We say that the representation ψ is *irreducible* if the representation module V belonging to ψ is an irreducible $F(G)$-module. Given two representations ψ and θ of G having representation modules V and W respectively, we say that they are *equivalent* if V and W are isomorphic as $F(G)$-modules. This states the following: the diagram

$$
\begin{array}{ccc}
V & \xrightarrow[\approx]{P} & W \\
\psi(g) \downarrow & & \downarrow \theta(g) \\
V & \xrightarrow[\approx]{P} & W
\end{array}
$$

is commutative for all $g \in G$. Writing out what this says we have that $P\theta(g) = \psi(g)P$ for all $g \in G$ where P is a vector space isomorphism of V onto W, or equivalently, $\theta(g) = P^{-1}\psi(g)P$ for all $g \in G$. This is an equivalence relation; most often when we speak of a representation we shall be speaking about the equivalence class itself.

DEFINITION. *If ψ is a representation of G then the character of ψ, χ_ψ, is defined by $\chi_\psi(g) = \text{tr}\,\psi(g)$.*

When there is no danger of confusion we shall often write χ_ψ merely as χ. Note that χ_ψ does not depend on ψ but rather on the equivalence class of ψ, for if ψ and θ are equivalent then $\theta(g) = P^{-1}\psi(g)P$ hence

$$\chi_\theta(g) = \text{tr}\,\theta(g) = \text{tr}\,P^{-1}\psi(g)P = \text{tr}\,\psi(g) = \chi_\psi(g).$$

Note also that χ is a *class function* in the sense that $\chi(g) = \chi(x^{-1}gx)$ for any x, $g \in G$. For

$$\psi(x^{-1}gx) = \psi(x)^{-1}\psi(g)\psi(x)$$

which yields, on taking traces,

$$\chi(x^{-1}gx) = \operatorname{tr} \psi(x)^{-1}\psi(g)\psi(x) = \operatorname{tr} \psi(g) = \chi(g).$$

In Lemma 4.3.1 we showed that if R is a semi-simple Artinian ring then any R-module is a direct sum of irreducible R-modules. Since $F(G)$ is a semi-simple Artinian ring (algebra) any $F(G)$-module is a direct sum of irreducible ones. If V is an $F(G)$-module and $V = \oplus V_i$, V_i irreducible $F(G)$-module and if ψ is the representation of G associated with V and ψ_i such that associated with V_i we write $\psi = \psi_1 \oplus \cdots \oplus \psi_m + \cdots$. We call the ψ_i's the *irreducible constituents* of ψ. If m_i denotes the number of V_j in this decomposition which are module isomorphic to V_i then, when V is finite-dimensional and so all the m_i are finite, we symbolically write $\psi = \sum m_i\psi_i$ and call m_i the *multiplicity* of ψ_i in ψ. In this situation, taking traces yields $\chi = \sum m_i\chi_i$ where χ is the character of ψ and χ_i that of ψ_i. By and large we shall be most concerned with the irreducible representations and characters of G.

In Lemma 4.3.2 we showed that any irreducible R-module, for R a semi-simple Artinian ring, is isomorphic to a minimal right ideal of R. Thus this holds true for $F(G)$. Now we know that $F(G) \approx F_{n_1} \oplus \cdots \oplus F_{n_k}$ where F_{n_i} is the ring of $n_i \times n_i$ matrices over F. The minimal right ideals of $F(G)$ are those of the various F_{n_i}. In F_{n_i} the minimal right ideals are isomorphic to

$$\rho_1^{(i)} = \left\{ \begin{pmatrix} \alpha_1 & \alpha_2 & \cdots & \alpha_{n_i} \\ 0 & 0 & \cdots & 0 \\ \vdots & \vdots & & \vdots \\ 0 & 0 & \cdots & 0 \end{pmatrix} \,\middle|\, \alpha_j \in F \right\}$$

and F_{n_i} is a direct sum of n_i isomorphs of $\rho_1^{(i)}$. Hence G only has a finite number of inequivalent irreducible representations, in fact at most k such. We claim that the k representations afforded us by the minimal right ideals of $\rho_1^{(1)}, \cdots, \rho_1^{(k)}$ of $F_{n_1}, \cdots, F_{n_k}$ respectively are indeed inequivalent. For $F_{n_i} = e_i F(G)$ where e_i is a central idempotent in $F(G)$ and where $e_i e_j = 0$ for $i \neq j$. Hence if $i \neq j$ since $\rho_1^{(i)} e_i = \rho_1^{(i)}$ and $\rho_1^{(j)} e_i = (0)$, $\rho_1^{(i)}$ and $\rho_1^{(j)}$ cannot be isomorphic as $F(G)$-modules. Hence G has exactly k distinct irreducible representations. Recall that $k = \dim {}_F Z(F(G))$ where $Z(F(G))$ is the center of $F(G)$. We shall need this remark later. If ψ_i denotes the irreducible representation having $\rho_1^{(i)}$ as representation module then we call n_i the *degree* of ψ_i.

Let us recall that the right regular representation τ of $F(G)$ is defined by $\tau(a) = T_a$ where $xT_a = xa$ for all $x \in F(G)$. In other words $F(G)$ itself serves as the representation module for τ. Since $F(G) = F_{n_1} \oplus \cdots \oplus F_{n_k}$ and each F_{n_i} is a direct sum of n_i isomorphic copies of $\rho_1^{(i)}$ we have the

THEOREM 5.1.1. $\tau = \sum n_i \psi_i$; *that is, every irreducible representation of G is a constituent of the regular representation as often as its degree.*

The theorem has two immediate and interesting consequences.

COROLLARY 1. $o(G) = \sum_{i=1}^k n_i^2$.

Proof. $o(G) = \dim F(G) = \sum \dim F_{n_i} = \sum n_i^2$.

COROLLARY 2. *If $g \neq 1$ is in G then* $\chi_\tau(g) = \sum n_i \chi_i(g) = 0$.

Proof. Using the group elements of G as a basis for $F(G)$ we have that, since for $g \neq 1$ $g_i T_g \neq g_i$, on the diagonal of T_g as a matrix in this basis we only have 0's.

Therefore $\chi_\tau(g) = \operatorname{tr} T_g = 0$. Since $\chi_\tau = \sum n_i \chi_i$ we get the full result.

Note that the representation of G, which we write as ψ_1, sending each $g \in G$ into 1 is an irreducible representation of G of degree 1. Hence $n_1 = 1$. We call ψ_1 the *unit representation* and χ_1 the *unit character* of G. If ψ_i is an irreducible representation of degree 1 we call it a *linear representation* of G. How many such linear representations does G have? The answer is provided in

THEOREM 5.1.2. *If G' is the commutator subgroup of G then the number of linear representations of G equals $o(G/G')$.*

Proof. The group G/G' is abelian hence $F(G/G')$ is a commutative, semi-simple algebra, thus its simple constituents are fields and hence are isomorphic to F. In other words all the irreducible representations of G/G' are linear. If $\bar\theta$ is an irreducible representation of G/G' we define a representation θ of G by means of $\theta(g) = \bar\theta(gG')$. We leave it to the reader to verify that this is a linear representation of G. Hence all the $o(G/G')$ distinct linear representations of G/G' induce distinct linear representations of G.

On the other hand if θ is a linear representation of G then $\theta(G)$, as a subgroup of F, is abelian; thus $G' \subset \operatorname{Ker} \theta$. We define a representation $\bar\theta$ on G/G' by $\bar\theta(gG') = \theta(g)$ for all $g \in G$. Since $G' \subset \operatorname{Ker} \theta$ this is well defined and is a linear representation of G/G'. Hence G has at most $o(G/G')$ distinct linear representations. Combined with what we did above this yields the result.

Note that in the course of the proof we actually proved a little more, namely: if H is a homomorphic image of G then any irreducible representation of H induces one for G and, conversely, any representation of

G having Ker $(G{\to}H)$ in its kernel defines a representation on H.

We return to the question of the number of distinct, inequivalent, irreducible representations that G possesses.

THEOREM 5.1.3. *The number of distinct inequivalent, irreducible representations of G equals the number of distinct conjugate classes in G.*

Proof. As we pointed out earlier, the number of distinct inequivalent irreducible representations of G equals dim $_F Z(F(G))$. We calculate dim $_F(Z(F(G)))$ another way, making heavy use of the fact that $F(G)$ has a very special basis.

If $g{\in}G$ let $C(g)$ be the conjugate class of g in G and let $C_g = \sum_{x\in C(g)} x$. We call C_g the *class sum* of g. Clearly C_g commutes with all the elements of G hence with all the elements in $F(G)$, that is, $C_g{\in}Z(F(G))$. Since the group elements are linearly independent over F the C_g's are also linearly independent over F. We claim that the C_g's constitute a basis of $Z(F(G))$ over F. Let $z = \sum\alpha_i g_i$ be in $Z(F(G))$ where the $\alpha_i{\in}F$, $g_i{\in}G$. If $x{\in}G$ then $\sum\alpha_i g_i = z = xzx^{-1} = \sum\alpha_i x g_i x^{-1}$; since the group elements are linearly independent over F we have, from the comparison of coefficients of the first and last terms of this equality, that every conjugate of g_i enters the expression of z with the same coefficient as does g_i. Therefore $z = \sum\alpha_i C_{g_i}$, in consequence of which the C_g's form a basis of $Z(F(G))$. Thus dim $_F(Z(F(G)))$ equals the number of conjugate classes in G.

Although our main concern is with the group algebra of a finite group over the field of complex numbers we now go on to find the analog of Theorem 5.1.3 for the group algebra of a finite group over any field, algebraically closed but of arbitrary characteristic. The result and method came from a paper of Richard Brauer. The next lemma, aside from its importance to the ultimate result we seek, has an interest all its own.

LEMMA 5.1.1. *Let A be an algebra finite dimensional over an algebraically closed field E of characteristic $p \neq 0$. Let S be the vector space in A spanned by all $ab - ba$, a, $b \in A$, over F and let $T = \{x \in A \mid x^{p^n} \in S \text{ for some } n\}$. Then T is a subspace of A, and the number of simple constituents of $A/J(A)$ equals the dimension of A/T as a vector space over E.*

Proof. If a, $b \in A$ then expanding formally and combining terms using cyclic permutations we see that $(a+b)^p \equiv a^p + b^p \bmod S$. Hence

$$(ab - ba)^p \equiv (ab)^p - (ba)^p \equiv a\{(ba)^{p-1}b\}$$

$$- \{(ba)^{p-1}b\}a \equiv 0 \bmod S.$$

In other words pth powers of elements of S fall back in S, hence $S \subset T$. If a, $b \in T$ then since $(a+b)^{p^k} \equiv a^{p^k} + b^{p^k}$ mod S we immediately see that T is a subspace of A.

If A is simple then by Wedderburn's theorem and the algebraic closure of E we have that $A \approx E_n$; moreover, in this case $S = \{a \in E_n \mid \operatorname{tr} a = 0\}$. Hence $\dim A/S = 1$; since $A \supset T \supset S$ and $A \neq T$ for

$$\begin{pmatrix} 1 & 0 & \cdots & 0 \\ 0 & 0 & \cdots & 0 \\ \cdot & \cdot & \cdots & \cdot \\ 0 & 0 & \cdots & 0 \end{pmatrix} \in A$$

and is not in T we get that $S = T$ and $\dim A/T = 1$.

In general, since $J(A)$ is nilpotent it must be in T so in $A/J(A) = \overline{A}$ we have that $\overline{T} = T/J(A)$ and $\dim A/T = \dim \overline{A}/\overline{T}$. But since $\overline{A}$ is a direct sum of simple algebras $\overline{A}_i$ and $\overline{T}$ is the direct sum of the corresponding $\overline{T}_i$, knowing that each dim $\overline{A}_i/\overline{T}_i = 1$ gives us that dim $\overline{A}/\overline{T}$ is the number of simple constituents of $\overline{A}$. From dim $A/T = \dim \overline{A}/\overline{T}$ the result now follows.

DEFINITION. *If $g \in G$ then g is said to be p-regular, p a prime, if the order of g is not divisible by p.*

If $p \nmid o(G)$ then, of course, every element of G is p-regular. We now prove the extended form of Theorem 5.1.3.

THEOREM 5.1.4. *Let G be a finite group and let E be an algebraically closed field of characteristic $p \neq 0$. Then the number of distinct inequivalent irreducible representations of $E(G)$ equals the number of conjugate classes of p-regular elements in G.*

Proof. Let $A = E(G)$ and let S and T have the same meaning they had in the last lemma. Since by definition of the radical $J(A) \subset \operatorname{Ker} \psi$ for any irreducible representation ψ of A, the irreducible representations of A are merely those of $A/J(A)$.

Given $g \in G$, from the Sylow decomposition of the cyclic group of G generated by g we get that $g = ab = ba$ where a is p-regular and b has order p^k. Now

$$(ab - a)^{p^k} \equiv a^{p^k} b^{p^k} - a^{p^k} \equiv a^{p^k} - a^{p^k} \equiv 0 \bmod S,$$

hence $ab - a \in T$ and so $g \equiv a \bmod T$. In other words, mod T every element of G is congruent to a p-regular element. Also if $g_1, g_2 \in G$ are conjugate then from

$$g_2 = x g_1 x^{-1} = x(g_1 x^{-1}) - (g_1 x^{-1})x + g_1$$

we get $g_2 \equiv g_1 \bmod S$, hence $g_2 \equiv g_1 \bmod T$. In consequence of all this we get that, given $x \in A$, then $x \equiv \sum \alpha_i a_i$ mod T where the a_i are representatives of the p-regular classes. Thus these span A/T. We show that they are linearly independent modulo T.

Suppose that $x \equiv \sum \alpha_i a_i \equiv 0 \bmod T$, where the a_i represent the distinct p-regular conjugate classes. We can pick $q = p^k$ large enough so that both $x^q \in S$ and each $a_i{}^q = a_i$ (since the a_i are p-regular). Hence

$$x^q \equiv \sum \alpha_i{}^q a_i{}^q \equiv \sum \alpha_i{}^q a_i \equiv 0$$

mod S. But if $x^q \in S$ then as is easily verified the sum of its coefficients over each conjugate class is 0. This forces $\alpha_i{}^q = 0$ and so $\alpha_i = 0$. In other words the p-regular class representatives a_i form a basis of A modulo T. Thus, using the preceding lemma, from dim $A/T =$ number of p-regular classes, we get the theorem.

Note that this result has Theorem 2.3.2 as a special case—at least for algebraically closed fields—for if $o(G) = p^n$ then G has only one p-regular class, namely $\{1\}$.

We return to the representation theory of finite groups over the field of complex numbers. Let χ be the character of a representation ψ of G. Since for $g \in G$,

$$(\psi(g))^{o(G)} = \psi(g^{o(G)}) = \psi(1) = I,$$

the characteristic roots of $\psi(g)$ are roots of unity. Thus $\chi(g) = \operatorname{tr} \psi(g)$ is a sum of the characteristic roots of $\psi(g)$ and so, a sum of roots of unity. Therefore $\chi(g)$ is an algebraic integer. We have proved

LEMMA 5.1.2. *If χ is the character of a representation of G then, for all $g \in G$, $\chi(g)$ is an algebraic integer.*

Let ψ_i be an irreducible representation of G of degree n_i; then $\psi_i(C_g)$ is in the center of F_{n_i} so is a scalar, that is, $\psi_i(C_g) = \omega_i(g) I_i$ where $\omega_i(g) \in F$ and where I_i is the $n_i \times n_i$ unit matrix. Taking traces and using that χ_i is a class function we obtain $h_g \chi_i(g) = n_i \omega_i(g)$ where h_g denotes the number of conjugates of g in G. Hence $\omega_i(g) = (h_g \chi_i(g))/n_i$.

Our immediate goal now is to establish that $\omega_i(g)$ is an algebraic integer for all $g \in G$ and all $i = 1, 2, \cdots, k$.

THEOREM 5.1.5. *If ψ_i is an irreducible representation of G then $\psi_i(C_g) = \omega_i(g)I_i$ where $\omega_i(g)$ is an algebraic integer.*

Proof. We have shown everything except that $\omega_i(g)$ is an algebraic integer.

If $a, b \in G$ then $C_aC_b = \sum \gamma_{abg}C_g$ where $\gamma_{abg} \geqq 0$ is an integer. To be precise γ_{abg} is the number of times $g = a'b'$ where a' is a conjugate of a and b' one of b. Applying ψ_i to this we get

$$\psi_i(C_aC_b) = \sum \gamma_{abg}\psi_i(C_g)$$

and so

$$\psi_i(C_a)\psi_i(C_b) = \sum \gamma_{abg}\psi_i(C_g).$$

Using the definition of ω_i we get

$$\omega_i(a)\omega_i(b) = \sum \gamma_{abg}\omega_i(g).$$

Let a be fixed in this but let b vary over the distinct conjugate classes, taking on the values $b_1, \cdots, b_k$. The system of equations becomes:

$$(\omega_i(a) - \gamma_{ab_1b_1})\omega_i(b_1) - \gamma_{ab_1b_2}\omega_i(b_2) - \cdots - \gamma_{ab_1b_k}\omega_i(b_k) = 0$$
$$\vdots \qquad \qquad \vdots \qquad \qquad \vdots$$
$$-\gamma_{ab_kb_1}\omega_i(b_1) - \cdots \cdots + (\omega_i(a) - \gamma_{ab_kb_k})\omega_i(b_k) = 0.$$

Since $\omega_i(1) = 1$ not all the $\omega_i(b_j)$ are 0; since the above system has a nontrivial solution we must have

$$\begin{vmatrix} (\omega_i(a) - \gamma_{ab_1b_1}) & -\gamma_{ab_1b_2} & \cdots & -\gamma_{ab_1b_k} \\ -\gamma_{ab_2b_1} & (\omega_i(a) - \gamma_{ab_2b_2}) & \cdots & -\gamma_{ab_2b_k} \\ \vdots & \vdots & & \vdots \\ -\gamma_{ab_kb_1} & -\gamma_{ab_kb_2} & \cdots & (\omega_i(a) - \gamma_{ab_kb_k}) \end{vmatrix} = 0.$$

In consequence $\omega_i(a)$ is the root of a monic polynomial with integer coefficients. In short, $\omega_i(a)$ is an algebraic integer.

COROLLARY. *For all* $g \in G$, $(h_\sigma \chi_i(g))/n_i$ *is an algebraic integer.*

Proof. As we saw before the proof of the theorem, $\omega_i(g) = (h_\sigma \chi_i(g))/n_i$.

We compute the average of a character over G; the result we get is a special case of the orthogonality relations we prove later.

LEMMA 5.1.3. *If* χ *is the character of an irreducible representation* ψ *of* G, *where* $\psi \neq \psi_1$, *then* $\sum_{g \in G} \chi(g) = 0$.

Proof. Consider the element $x = \sum_{g \in G} g$ in $F(G)$. If $a \in G$ then $xa = ax = x$. In particular $x \in Z(F(G))$. Since $\psi \neq \psi_1$ there is an $a \in G$ such that $\psi(a) \neq I$.

Now $\psi(xa) = \psi(x)$ which yields that $\psi(x)\psi(a) = \psi(x)$. Since $x \in Z(F(G))$, $\psi(x) = \lambda I$ for some $\lambda \in F$ hence we get $\lambda\psi(a) = \lambda I$. As $\psi(a) \neq I$ we get $\lambda = 0$, that is, $\psi(x) = 0$. Taking traces we get $\sum_{g \in G} \chi(g) = 0$.

If ψ is an irreducible representation of G we can consider $\psi(g)$ as a matrix over F. We define $\psi^*(g) = $ transpose of $\psi(g^{-1}) = \psi(g^{-1})'$. It is immediate that ψ^* is an irreducible representation of G. We call ψ^* the *contragredient representation* of ψ. What is χ^*, the character of ψ^*? If α is a characteristic root of $\psi(g)$ then α^{-1} is a characteristic root of $\psi(g^{-1})$ and so of $\psi^*(g) = \psi(g^{-1})'$. Moreover the multiplicity of α as a characteristic root of $\psi(g)$ equals that of α^{-1} as a characteristic root of $\psi^*(g)$. Now, as noted earlier, α is a root of unity hence $\alpha^{-1} = \bar{\alpha}$, the complex conjugate of α. All in all we have that $\chi^*(g) = \overline{\chi(g)}$.

Let ψ_i and ψ_j be irreducible representations of G with characters χ_i and χ_j respectively. We define $\psi_i \otimes \psi_j$ by

$(\psi_i \otimes \psi_j)(g) = \psi_i(g) \otimes \psi_j(g)$ for $g \in G$. It is clear that this defines a representation of G. Its character, moreover, from elementary properties of the tensor (or Kronecker) product of linear transformations, is $\chi_i \chi_j$. As a representation of G, $\psi_i \otimes \psi_j$ can be written as a direct sum of irreducible representations, hence

$$\psi_i \otimes \psi_j = \sum_{q=1}^{k} t_{ijq} \psi_q$$

where the $t_{ijq} \geqq 0$ are integers. Computing traces we obtain

$$\chi_i(g) \chi_j(g) = \sum_{q} t_{ijq} \chi_q(g).$$

Can we predict what constituents will be present in the decomposition of $\psi_i \otimes \psi_j$; otherwise put, when is $t_{ijq} \neq 0$? Fixing on the unit representation, a less ambitious question might be: when is ψ_1 a constituent of $\psi_i \otimes \psi_j$, that is, when is $t_{ij1} \neq 0$? To answer this we first consider the converse question.

LEMMA 5.1.4. *If ψ_i and ψ_j are irreducible representations of G such that the unit representation of G is a constituent of $\psi_i \otimes \psi_j$ then ψ_i is equivalent to ψ_j^*.*

Proof. Let V be a representation module for ψ_i and W one for ψ_j. Then $V \otimes W$ serves as a representation module for $\psi_i \otimes \psi_j$.

Since the unit representation is a constituent of $\psi_i \otimes \psi_j$ there must be a vector $z \neq 0$ in $V \otimes W$ such that

$$z(\psi_i(g) \otimes \psi_j(g)) = z$$

for all $g \in G$.

Let $z = \sum v_\alpha \otimes w_\alpha$ where the w_α are in W and are linearly independent over F; we suppose that each v_α appearing is not 0. Let V_0 be the subspace of V spanned by these v_α's. We claim that $V_0\psi_i(G) \subset V_0$. To see this note that if $g \in G$ then

$$z = z(\psi_i(g) \otimes \psi_j(g)) = \sum v_\alpha\psi_i(g) \otimes w_\alpha\psi_j(g).$$

From the linear independence of the w_α's and the properties of the tensor product we get that each $v_\alpha\psi_i(g) \in V_0$. The irreducibility of $\psi_i(g)$ implies that $V_0 = V$. Similarly we get that the w_α must span W. Hence dim $V \leqq$ dim W and dim $W \leqq$ dim V yielding that dim $V =$ dim W. Note that the argument actually reveals a little more, namely that if $z = \sum v_\alpha \otimes w_\alpha$ and if the w_α are linearly independent over F they form a basis of W and the v_α must be linearly independent and form a basis of V.

Let $v_\alpha\psi_i(G) = \sum_\beta \mu_{\alpha\beta}v_\beta$ and $w_\alpha\psi_j(g) = \sum_\beta \nu_{\alpha\beta}w_\beta$. The matrices $(\mu_{\alpha\beta})$ and $(\nu_{\alpha\beta})$ are the matrices of $\psi_i(g)$ and $\psi_j(g)$ respectively in the bases indicated.

Since $z = \sum v_\alpha \otimes w_\alpha$ is left fixed by all the $\psi_i(g) \otimes \psi_j(g)$ we have that

$$\sum_\alpha \left(\sum_\beta \mu_{\alpha\beta}v_\beta \right) \otimes \left(\sum_\beta \nu_{\alpha\beta}w_\beta \right) = \sum_\beta v_\beta \otimes w_\beta;$$

the linear independence of the $v_\gamma \otimes w_\delta$ allows us to read off from the above relations:

$$(1) \quad \sum_\alpha \mu_{\alpha\beta}\nu_{\alpha\beta} = 1 \quad \text{for each } \beta$$

$$(2) \quad \sum_\alpha \mu_{\alpha\beta}\nu_{\alpha\gamma} = 0 \quad \text{for } \beta \neq \gamma.$$

We thus have that the matrices $(\mu_{\alpha\beta})$ and $(\nu_{\alpha\beta})$ are related by $(\mu_{\alpha\beta})(\nu_{\alpha\beta})' = I$. Thus $\psi_i(g)\psi_j(g)' = I$. We have proved that ψ_i must be equivalent to $\psi_j{}^*$.

The lemma has a very interesting consequence, namely

COROLLARY. *If ψ_i and ψ_j are irreducible representations of G and ψ_i is not equivalent to ψ_j^* then $\sum_{g \in G} \chi_i(g)\overline{\chi_j(g)} = 0$.*

Proof. Since ψ_i is not equivalent to ψ_j^*, by the lemma the unit representation is not a constituent of $\psi_i \otimes \psi_j^*$. Thus $\psi_i \otimes \psi_j^* = \sum \tilde{l}_{ijq}\psi_q$ where $\tilde{l}_{ij1} = 0$. Hence for $g \in G$, taking traces,

$$\chi_i(g)\overline{\chi_j(g)} = \sum_q \tilde{l}_{ijq}\chi_q(g).$$

Summing this over g and using Lemma 5.1.3 we get

$$\sum_{g \in G} \chi_i(g)\overline{\chi_j(g)} = \sum_q \tilde{l}_{ijq} \sum_{g \in G} \chi_q(g) = \tilde{l}_{ij1} \sum_{g \in G} \chi_1(g) = 0,$$

since $\tilde{l}_{ij1} = 0$.

If ψ_i is not equivalent to ψ_j^* we now know that ψ_1 is not a constituent of $\psi_i \otimes \psi_j^*$. But must it be a constituent of $\psi_i \otimes \psi_i^*$ for each i, and if so, what is its multiplicity?

LEMMA 5.1.5. *The unit representation ψ_1 is a constituent of $\psi_i \otimes \psi_i^*$ with multiplicity 1. Moreover $\sum_{g \in G} \chi_i(g)\overline{\chi_i(g)} = o(G)$.*

Proof. By Corollary 2 to Theorem 5.1.1 for $g \neq 1$ in G, $\sum_i n_i\chi_i(g) = 0$ and for $g = 1$, $\sum_i n_i\chi_i(1) = \sum n_i^2 = o(G)$.

If $g \neq 1$ then $\sum_i n_i\chi_i(g)\overline{\chi_j(g)} = 0$; we sum this over all $g \neq 1$ in G to get

$$(1) \qquad \sum_i n_i \sum_{g \neq 1} \chi_i(g)\overline{\chi_j(g)} = 0.$$

By the corollary to Lemma 5.1.4 for $\chi_i \neq \chi_j$ then $\sum_{g \in G} \chi_i(g)\overline{\chi_j(g)} = 0$, hence

$$\sum_{g \neq 1} \chi_i(g)\overline{\chi_j(g)} = -\chi_i(1)\overline{\chi_j(1)} = -n_i n_j.$$

In view of this (1) becomes:

$$\sum_{i \neq j} n_i(-n_i n_j) + n_j \sum_{g \neq 1} \chi_j(g)\overline{\chi_j(g)} = 0.$$

Therefore

$$\sum_{g \neq 1} \chi_j(g)\overline{\chi_j(g)} = \sum_{i \neq j} n_i^2.$$

Adding $n_j^2 = \chi_j(1)\overline{\chi_j(1)}$ to both sides yields

$$\sum_{g \in G} \chi_j(g)\overline{\chi_j(g)} = \sum_i n_i^2 = o(G)$$

on invoking Corollary 1 to Theorem 5.1.1.

Now $\psi_i \otimes \psi_i^* = \sum_q l_{iiq}\psi_q$ hence for $g \in G$, $\chi_i(g)\overline{\chi_i(g)}$ $= \sum l_{iiq}\chi_q(g)$. Summing this over all $g \in G$ and using Lemma 5.1.3 and the result just established above gives us

$$o(G) = l_{ii1} \sum_{g \in G} \chi_1(g) = l_{ii1}o(G).$$

Hence $l_{ii1} = 1$, the desired result.

By an irreducible character of G we mean the character of an irreducible representation of G. The two previous lemmas combine to give an important orthogonality relation for the irreducible characters of G.

THEOREM 5.1.6. *If χ_i and χ_j are irreducible characters of G then*

$$\sum_{g \in G} \chi_i(g)\overline{\chi_j(g)} = \delta_{ij}o(G)$$

where δ_{ij} is the Kronecker delta.

The orthogonality relation contained in the statement of Theorem 5.1.5 implies yet another orthogonality relation,

THEOREM 5.1.7. *If a, $b \in G$ then*

$$\sum_{i=1}^{k} \chi_i(a)\chi_i(b) = 0$$

if a and b are not conjugate in G and equals the order of the normalizer of a in G if they are conjugate.

Proof. Let $g_1, \cdots, g_k$ be representatives of the conjugate classes of G and let $\chi_1, \cdots, \chi_k$ be the distinct irreducible characters of G. Let $A = (a_{ij})$ be the $k \times k$ matrix defined by $a_{ij} = \sqrt{h_j}\chi_i(g_j)$ where h_j denotes the number of elements in the conjugate class of g_j. Now $A\overline{A}' = (b_{ij})$ where

$$b_{ij} = \sum_{\nu} a_{i\nu}\bar{a}_{j\nu} = \sum_{\nu} h_{\nu}\chi_i(g_{\nu})\overline{\chi_j(g_{\nu})}.$$

Since characters are class functions this last sum can be written as $\sum_{g \in G} \chi_i(g)\overline{\chi_j(g)}$; by Theorem 5.1.6 this is $\delta_{ij}o(G)$. In other words, $A\overline{A}' = o(G)I$. By elementary matrix theory we deduce that $\overline{A}'A = o(G)I$. Since the (i, j) entry of $\overline{A}'A$ is

$$\sum_{\nu} \bar{a}_{\nu i}a_{\nu j} = \sqrt{h_i}\sqrt{h_j}\sum_{\nu} \overline{\chi_{\nu}(g_i)}\chi_{\nu}(g_j)$$

we get that for $i \neq j$ (that is, $g_i \neq g_j$)

$$\sum_{\nu} \overline{\chi_{\nu}(g_i)}\chi_{\nu}(g_j) = 0$$

and for $g_i = g_j$,

$$h_i \sum_{\nu} \chi_{\nu}(g_i)\overline{\chi_{\nu}(g_i)} = o(G).$$

Since the number of elements in the conjugate class of g_i is $h_i = o(G)/o(N(g_i))$ where $N(g_i)$ is the normalizer of g_i in G we obtain the assertion of the theorem.

Let ψ be a representation of G; therefore $\psi = \sum_i m_i\psi_i$. $\chi(g) = \sum_i m_i\chi_i(g)$ for all $g \in G$. From this we immediately pass to

$$\sum_{g \in G} \chi(g)\overline{\chi_j(g)} = \sum_i m_i \sum_{g \in G} \chi_i(g)\overline{\chi_j(g)}$$

$$= \sum_i \delta_{ij}m_i o(G) = m_j o(G).$$

Consequently $m_j = 1/o(G) \sum_g \chi(g)\overline{\chi_j(g)}$. What we have shown is that from the knowledge of the character of ψ we can compute the multiplicity of each irreducible representation as a constituent of ψ. Knowing this we know ψ. Thus

THEOREM 5.1.8. *The character of a representation of G determines the representation; that is, two representations having the same character are equivalent.*

Since the k complex functions $\chi_1, \cdots, \chi_k$ are orthogonal on the set of conjugate classes which is a set consisting of k elements any complex valued function f on G which is constant on the conjugate classes can be written as $f = \sum \alpha_i \chi_i$ with $\alpha_i \in F$.

We close this section with another basic fact intertwining the irreducible representations of a group with the group itself.

THEOREM 5.1.9. *If n_i is the degree of an irreducible representation ψ_i of G then $n_i \mid o(G)$.*

Proof. Lemma 5.1.2 tells us that $\overline{\chi_i(g)}$ is an algebraic integer for any $g \in G$. The corollary to Theorem 5.1.5 assures us that $(h_g \chi_i(g))/n_i$ is an algebraic integer, hence $(h_g \chi_i(g)\overline{\chi_i(g)})/n_i$ is an algebraic integer for any $g \in G$. If $g_1, \cdots, g_k$ are representatives of the conjugate classes of G then

$$\sum_{j=1}^k \frac{h_j \chi_i(g_j)\overline{\chi_i(g_j)}}{n_i} = \sum_{g \in G} \frac{\chi_i(g)\overline{\chi_i(g)}}{n_i} = \frac{1}{n_i} \sum_{g \in G} \chi_i(g)\overline{\chi_i(g)}$$

is an algebraic integer. By Theorem 5.1.6 $\sum_{g \in G} \chi_i(g)\overline{\chi_i(g)}$ $= o(G)$ in consequence of which $o(G)/n_i$ is an algebraic integer. Being rational it must be a rational integer. Hence $n_i \mid o(G)$.

2. A theorem of Hurwitz. We shall use the results established about the representations of finite groups to obtain a lovely result of Hurwitz about the composition of quadratic forms.

Let F be a field—we suppose it is the complex field although the argument given would be valid as long as its characteristic is not 2 —and we ask: for what values of m and n is $(x_1^2 + \cdots + x_m^2)(y_1^2 + \cdots + y_n^2)$ $= z_1^2 + \cdots + z_n^2$ where the z_i are bilinear functions of the x's and y's? A special case, but one of great interest, is that in which $m = n$; the question becomes: when does the sum of n squares compose bilinearly? For $n = 1$, 2, 4 or 8 we can write down explicit formulas to show that the respective sums of squares compose. As a consequence of the theorem we shall establish it will turn out that these are the only cases.

The treatment we shall follow is due to Eckmann.

Suppose then that

$$(1) \quad (x_1^2 + \cdots + x_m^2)(y_1^2 + \cdots + y_n^2) = z_1^2 + \cdots + z_n^2$$

where the z_i are bilinear functions of the x's and y's. The case $m = 2$ is trivial, *so in what follows we suppose that $m > 2$.*

We write $z_i = \sum_{j=1}^n a_{ij}(x)y_j$ where the $a_{ij}(x)$ are linear functions in the x_k's. Substituting in (1), in equating coefficients, we get

$$(2) \quad \sum_i a_{ij}(x)a_{ik}(x) = 0 \quad \text{for } j \neq k$$

$$\sum_i a_{ij}^2(x) = x_1^2 + \cdots + x_m^2.$$

Thus if A denotes the $n \times n$ matrix $(a_{ij}(x))$ the relations (2) can be succinctly written as $AA' = (x_1^2 + \cdots + x_m^2)I$ where I is the identity matrix.

Because the $a_{ij}(x)$ are linear functions in the x's we can write A as $A = A_1x_1 + \cdots + A_mx_m$ where the A_i are $n \times n$ matrices over F. The equation $AA' = (x_1^2 + \cdots + x_m^2)I$ immediately yields

$$(3) \quad A_iA_j' + A_jA_i' = 0 \quad \text{for } i \neq j, A_iA_i' = I.$$

We have reduced the problem of the composition of the sums of squares to the following matrix question: when can we find m complex $n \times n$ matrices which satisfy (3)?

First of all we make a normalization; let $B_i = A_m'A_i$. In these terms (3) becomes:

$$(4) \quad B_iB_j' + B_jB_i' = 0 \quad \text{for } i \neq j, B_iB_i' = I, B_m = I.$$

If one puts $j = m$ in the first relation of (4) we get that $B_i' = -B_i$ for $i \neq m$. Thus we are asking: for what integers m and n can we find $m-1$ complex, skew-symmetric, orthogonal $n \times n$ matrices which anti-commute amongst themselves?

We leave the matrices for a moment. Let G be a finite group generated by the elements $a_1, \cdots, a_{m-1}, \epsilon$ subject to: $a_i^2 = \epsilon \neq 1$, $\epsilon^2 = 1$, $a_ia_j = \epsilon a_ja_i$ for $i \neq j$. Note that $a_i\epsilon = \epsilon a_i$ so that ϵ is in the center of G. Recall that we are assuming that $m > 2$. G is clearly a group of order 2^m whose commutator subgroup G' is $\{1, \epsilon\}$ of order 2.

We make some simple observations about G:

1. if m is odd the center of G consists merely of 1 and ϵ.
2. if m is even the center of G consists of the elements $1, \epsilon, a_1a_2 \cdots a_{m-1}, \epsilon a_1a_2 \cdots a_{m-1}$.
3. if $g \in G$ is not in the center its conjugate class consists of g and ϵg.

Therefore:

1. when m is odd G has $2+(2^m-2)/2 = 2^{m-1}+1$ conjugate classes.
2. when m is even G has $4+(2^m-4)/2 = 2^{m-1}+2$ conjugate classes.

By Theorem 5.1.2, since $o(G/G') = 2^{m-1}$, G has 2^{m-1} linear representations. The total number of distinct irreducible representations of G equals the number of conjugate classes in G. Thus when m is odd G has one irreducible representation of degree $f \neq 1$. By Corollary 1 to Theorem 5.1.1 the sum of the squares of the degrees of the irreducible representations of G is the order of G, hence $2^{m-1}+f^2 = 2^m$, leaving us with $f = 2^{(m-1)/2}$. When m is even, on the other hand, G has two irreducible representations of degree f_1, f_2 different from 1. Hence $f_1^2+f_2^2 +2^{m-1} = 2^m$. By Theorem 5.1.9 $f_i \mid o(G) = 2^m$ hence f_i is a power of 2. Combined with the above we get $f_1 = f_2$ and so $f_1 = f_2 = 2^{(m-2)/2}$.

We go back to our matrices; our question about them becomes for what n can we find a representation of G of degree n such that ϵ is represented by $-I$? Since ϵ is in G' it is represented by 1 in all the linear representations of G. Thus our representation must be a direct sum of the non-linear irreducible representations of G. This gives:

1. for m odd, $n = k2^{(m-1)/2}$
2. for m even, $n = k2^{(m-2)/2}$.

We have proved the

THEOREM 5.2.1. *If $n = 2^t s$ where s is odd then for $m > 2$ we can find m complex $n \times n$ matrices satisfying (4) if and only if $2^{(m-2)/2} \leq 2^t$, that is, if and only if $m \leq 2t+2$.*

A beautiful corollary to the result is the famous theorem of Hurwitz,

THEOREM 5.2.2. *If* $(x_1^2 + \cdots + x_n^2)(y_1^2 + \cdots + y_n^2)$ $= z_1^2 + \cdots + z_n^2$ *where the* z_i *are bilinear functions of the* x's *and* y's *then* $n = 1, 2, 4$ *or* 8.

The result, while proved in characteristic 0, actually holds in all characteristics other than 2 for the representation theory developed will hold as long as the characteristic of the field does not divide the order of the group. Since we were dealing with a group of order 2^m the only characteristic that could cause harm is 2.

3. Applications to group theory. We shall now make use of the theory of group characters and representations to obtain results about groups themselves. It is difficult to see or to explain why this machinery when turned loose on a group works so effectively, but work indeed it does.

We begin with a very easy little lemma.

LEMMA 5.3.1. *Let* ψ *be a representation of a finite group* G *with character* χ *and let* $N = \left\{ x \in G \mid |\chi(x)| = \chi(1) \right\}$. *Then* N *is a normal subgroup of* G *and in fact is* $\left\{ x \in G \mid \psi(x) = \alpha I, \ \alpha \in F \right\}$.

Proof. Let ψ be of degree n; then for $g \in G$, $\chi(g)$ is a sum of n roots of unity, namely the characteristic roots of $\psi(g)$. If $|\chi(g)| = \chi(1) = n$ then these roots of unity must all be equal, say to α. Hence $\psi(g) = \alpha I + a$ where a is a nilpotent matrix. However since $\psi(g)^{o(G)} = I$ and the characteristic of F is 0 we conclude that $a = 0$. Thus $\psi(g) = \alpha I$. Since we have shown that $N = \left\{ g \in G \mid \psi(g) = \alpha(g)I \text{ for some } \alpha(g) \in F \right\}$ it is clear that N is a normal subgroup of G.

The next lemma intertwines the nature of a representation—via its degree—with the local structure in the group—via its conjugate classes.

LEMMA 5.3.2. *If ψ_i is an irreducible representation of G of degree n_i and if $g \in G$ is such that h_g, the number of conjugates of g in G, is relatively prime to n_i then either $\chi_i(g) = 0$ or $\psi_i(g)$ is a scalar matrix.*

Proof. By the corollary to Theorem 5.1.5 $(h_g \chi_i(g))/n_i$ is an algebraic integer. However, from $(h_g, n_i) = 1$ we then deduce that $(\chi_i(g))/n_i$ is an algebraic integer. Now $\chi_i(g)$ is a sum of n_i roots of unity so that

$$\left| \frac{\chi_i(g)}{n_i} \right| \leqq 1.$$

If $|\chi_i(g)| = n_i = \chi_i(1)$ we know that $\psi_i(g)$ is a scalar matrix by the previous lemma. So we may suppose that $|\chi_i(g)/n_i| < 1$.

If β is an algebraic conjugate of $\chi_i(g)/n_i$ it is also of the form $1/n_i$ (sum of n_i roots of unity) in consequence of which $|\beta| \leqq 1$. Thus the product γ of all the algebraic conjugates of $(\chi_i(g))/n_i$ is *less* than 1 in absolute value. However, γ is an algebraic integer and, as the product of all the algebraic conjugates of $(\chi_i(g))/n_i$, γ is also rational. Therefore γ is a rational integer. In conjunction with $|\gamma| < 1$ we are forced to conclude that $\gamma = 0$ and so, $\chi_i(g) = 0$.

We now have all the pieces to prove a fundamental nonsimplicity criterion due to Burnside.

THEOREM 5.3.1. *If in G the number of conjugates of some elements $g \neq 1$ is a power of a prime then G cannot be simple.*

Proof. Suppose that for $g \neq 1$, $h_g = p^\alpha$ where p is a prime. By Theorem 5.1.1, Corollary 2, $\sum_i n_i \chi_i(g) = 0$ and since $n_1 = 1$ we have $1 + \sum_{i \neq 1} n_i \chi_i(g) = 0$. If G is simple $\psi_i(g)$ cannot be a scalar matrix hence if $p \nmid n_i$ then by the preceding lemma $\chi_i(g) = 0$. Therefore the only nonzero contributions to the sum $\sum_{i\neq1} n_i\chi_i(g)$ came from those n_i for which $p\,|\,n_i$. But since $\chi_i(g)$ is an algebraic integer we then get $1+p$ (algebraic integer) $=0$ which is manifestly impossible. The theorem is proved.

The theorem is both fundamental and interesting, but even more interesting perhaps is a simple consequence thereof which is a magnificent theorem due to Burnside.

THEOREM 5.3.2 (Burnside). *If $o(G)=p^mq^m$, p, q prime numbers then G is solvable.*

Proof. A standard result in group theory asserts that if N is a normal subgroup of G such that both N and G/N are solvable then G is solvable. In view of this, by use of induction, it suffices to show that G is not simple.

If $g\neq1$ is in the center of a q-Sylow subgroup then the normalizer of g contains a q-Sylow subgroup hence h_g, as the index of the normalizer of g in G, is of the form $h_g=p^\alpha$ for some α. By Theorem 5.3.1, G cannot be simple. This proves the result.

We close this chapter with another famous classical result in group theory; this one due to Frobenius. Both this theorem and the previous one, although purely group-theoretic in nature, have never been proved without recourse to the theory of group characters.

THEOREM 5.3.3 (Frobenius). *Let G be a finite group and suppose that H is a subgroup of G such that $x^{-1}Hx$ $\cap H=\{1\}$ if $x\notin H$. Then $N=G-\bigcup_{x\in G} x(H-1)x^{-1}$ is a normal subgroup of G. Also $G=HN$.*

To avoid any misunderstanding we point out that the "$-$" in the statement of the theorem is the set-theoretic minus.

Proof. The proof we give is due to Wielandt; he used it to prove a more general result.

Let $\phi \neq 1$ be an irreducible character of H. We use it to define a function θ on G as follows:

1.　　　　$\theta(g) = \phi(1) = t$　　　for $g \in N$.

2.　　　　$\theta(g) = \phi(x^{-1}gx)$　　　for $g \in xHx^{-1}$.

It is immediate from its definition that θ is a class function on G. The function Ω defined by $\Omega(g) = \theta(g) - t$ for all $g \in G$ is also a class function. Thus $\Omega = \sum \alpha_i \chi_i$ where the χ_i are the irreducible characters of G and where the α_i are in F. We compute the α_i. Now $\alpha_i = 1/o(G)$ $\cdot \sum_{g \in G} \Omega(g) \overline{\chi_i(g)}$. Since Ω vanishes on N and since the sum over any conjugate of H equals that over H this expression for α_i simplifies to $\alpha_i = (1/o(G))w \sum_{h \in H} \Omega(h)$ $\cdot \chi_i(h)$ where w is the number of conjugates of H in G. The hypothesis of the theorem forces H to be its own normalizer in G hence $w = o(G)/o(H)$ in light of which

$$\alpha_i = \frac{1}{o(H)} \sum_{h \in H} \Omega(h)\overline{\chi_i(h)} = \frac{1}{o(H)} \sum_{h \in H} (\theta(h) - t)\overline{\chi_i(h)}$$

$$= \frac{1}{o(H)} \sum_{h \in H} (\phi(h) - t)\overline{\chi_i(h)}.$$

Since ψ_i, the representation of which χ_i is the character, induces a representation on H by restriction we get from the orthogonality relation of characters on H that α_i is an integer for every i. What is α_i? A direct calculation shows that

$$\alpha_1 = \frac{1}{o(H)} \sum_{h \in H} (\phi(h) - t) = \frac{1}{o(H)} \sum_{h \in H} \phi(h) - t;$$

since $\phi \neq 1$ is an irreducible character of H, by Lemma 5.1.3 $\sum_{h \in H} \phi(h) = 0$ leaving us with $\alpha_1 = -t$.

Now $\sum_{g \in G} \Omega(g)\overline{\Omega(g)} = (\sum_i \alpha_i^2)o(G)$ by the orthogonality relation of the group characters. However, as above,

$$
\begin{aligned}
\sum_{g \in G} \Omega(g)\overline{\Omega(g)} &= \frac{o(G)}{o(H)} \sum_{h \in H} \Omega(h)\overline{\Omega(h)} \\
&= \frac{o(G)}{o(H)} \sum_{h \in H} (\phi(h) - t)\overline{(\phi(h) - t)} \\
&= \frac{o(G)}{o(H)} \left\{ \sum_{h \in H} \phi(h)\overline{\phi(h)} - t \sum_{h \in H} \phi(h) \right. \\
&\qquad\qquad \left. - t \sum_{h \in H} \overline{\phi(h)} + \sum_{h \in H} t^2 \right\} \\
&= \frac{o(G)}{o(H)} (o(H) + o(H)t^2),
\end{aligned}
$$

again by use of the orthogonality relations of the characters on H. Thus $o(G) \sum \alpha_i{}^2 = (t^2+1)o(G)$, yielding that $t^2+1 = \sum \alpha_i{}^2$. However $\alpha_1 = -t$ and the α_i are integers. Thus all but one of the a_i for $i \neq 1$ are 0 and that exceptional one is ± 1. Hence $\Omega = -t \pm \chi_i$ where χ_i is an irreducible character of G. For $g \in N$ since $\Omega(g) = 0$ we get that $\chi_i(g) = \pm t$ hence $|\chi_i(g)| = \chi_i(1)$, hence $N \subset \{g \in G \mid |\chi_i(g)| = \chi_i(1)\}$, a normal subgroup of G. Of course here the i is a function of the ϕ we started with.

As we let ϕ run over the nonunit characters of H we claim that N is the intersection of the groups $\{g \in G \mid |\chi_i(g)| = \chi_i(1)\}$ obtained; we leave this to the reader with the hint, if $g \neq 1 \in H$ pick an irreducible character ϕ of H such that $\phi(g) \neq \phi(1)$. Hence we see that N is a normal subgroup of G. That $G = HN$ is trivial by counting.

It was a long standing conjecture that the N in Theorem 5.3.3 is nilpotent. This was settled in the affirmative by John Thompson.

References

1. R. Brauer, Zur Darstellungstheorie der Gruppen endlicher Ordnung, *Math. Z.*, 63 (1956) 406–444.

2. C. Curtis and I. Reiner, *Representation Theory of Finite Groups*, Interscience, New York, 1962.

3. Beno Eckmann, Gruppentheoretischer Beweis des Satzes von Hurwitz-Radon über die Komposition quadratischer Formen, *Comment. Helv.*, 15 (1942) 358–366.

4. M. Hall, Jr., *The Theory of Groups*, Macmillan, New York, 1961.

5. I. N. Herstein, *Theory of Rings*, Univ. of Chicago, Math. Notes, 1961.

6. John Thompson, Finite groups with fixed-point-free automorphisms of prime order, *Proc. Nat. Acad. Sci.*, 45 (1959) 578–581.

7. H. Wielandt, "Über die Existenz von Normalteilern in endlichen Gruppen," *Math. Nachr.*, 18 (1958) 274–280.

CHAPTER 6

POLYNOMIAL IDENTITIES

As we have indicated earlier on several occasions the general structure theory developed in the first two chapters reveals itself most effectively in the study of rings which are further restricted by some sort of polynomial condition. A good instance of this was seen in the study of the commutativity of rings.

We now pass to a class of rings—in some sense they satisfy a higher form of commutativity—which on analysis are capable of a sharp description. These rings are subjected to the presence of some nonzero polynomial relation in noncommuting variables, which vanishes identically on them. The entry into their study is via an incisive result due to Kaplansky (Theorem 6.3.1).

This area has been studied from many points of view and with many diverse goals. For instance, much work has been done by Amitsur and Levitzki in investigating the nature of the minimal identities satisfied by certain well known rings. Our goal is in another direction; we seek to develop enough of the material to be able to settle the Kurosh problem for this family of objects.

1. A result on radicals. We begin the discussion with a result which is needed for the study to be made but which in spirit is very little related to it. Let R be a ring and let $R[t]$ be the ring of polynomials in an indeterminate t over R; we suppose that $(at)b = (ab)t$ for a, $b \in R$, that is, in case R has a unit element, t is in the center of $R[t]$.

THEOREM 6.1.1. *If R has no nonzero nil ideals then $R[t]$ is semi-simple.*

In order to carry out the proof we need two familiar and easy results. The first of these has actually been already proved in a larger setting—no commutativity being assumed—in Lemma 2.2.3.

LEMMA 6.1.1. *Let R be a commutative ring and let N be the intersection of all its prime ideals. Then N is a nil ideal.*

Proof. If $a \in R$ is not nilpotent we shall exhibit a prime ideal of R which excludes a. This will establish the lemma.

Let P be an ideal of R maximal with respect to the exclusion of $\{a^n \mid n = 1, 2, \cdots\}$. By Zorn's lemma such a P exists. We assert that P is a prime ideal of R. If $xy \in P$ and both x and y are not in P then since both $P + Rx$ and $P + Ry$ are properly larger than P we must have $a^m \in P + Rx$, $a^n \in P + Ry$ and so $a^{m+n} \in (P + Rx) \cdot (P + Ry) \subset P + Rxy \subset P$. This contradiction shows that P is prime.

With this in hand we are able to prove

LEMMA 6.1.2. *Let R be a commutative ring with unit element; if $a_0 + a_1 t + \cdots + a_n t^n \in R[t]$ is invertible in $R[t]$ then a_0 is invertible in R and $a_1, \cdots, a_n$ are nilpotent.*

Proof. Merely exploiting the definition of $R[t]$ and the invertibility of $a_0 + a_1 t + \cdots + a_n t^n$ leads us to conclude that a_0 is invertible in R. Now to the nilpotence of $a_1, \cdots, a_n$.

Let P be a prime ideal of R; then $P[t]$ is an ideal of $R[t]$ and, moreover,

$$\frac{R[t]}{P[t]} \approx R/P \, [t].$$

Since R/P is an integral domain the only units in $R/P[t]$ are the units of R/P; because $a_0 + a_1 t + \cdots + a_n t$ maps into such a unit we conclude that $a_i \equiv 0(P)$ for $i > 0$. Hence $a_1, \cdots, a_n$ are in all prime ideal of R which, by means of Lemma 6.1.1, tells us that they are nilpotent.

We are now ready to complete the proof of Theorem 6.1.1.

Let J be the radical of $R[t]$ and suppose that $J \neq (0)$. Let

$$0 \neq r = a_0 t^{n_0} + a_1 t^{n_1} + \cdots + a_k t^{n_k}, \; n_0 < n_1 < \cdots < n_k$$

be a nonzero element of J with as few nonzero coefficients $a_0, a_1, \cdots, a_k$ as possible. Now $a_i r - r a_i$ is in J and has fewer nonzero coefficients than r hence must be 0. From this we conclude that $a_i a_j = a_j a_i$ for all i and j.

Now rt is also in J hence, for some $s \in J$, $rt + s + rst = 0$. Therefore

$$s = -rt - rst = -rt - r(-rt - rst)t$$
$$= -rt + r^2 t^2 + r^2 s t^2.$$

Combining in this manner we get that for any n

$$s = -rt + r^2 t^2 - r^3 t^3 + \cdots$$
$$+ (-1)^n r^n t^n + (-1)^n r^n s t^n.$$

Pick n large enough to exceed the degree of s as a polynomial in t; then the term $r^n s t^n$ cannot contribute anything to the coefficients of s. In consequence the coefficients of s are polynomials in $1, a_0, \cdots, a_k$. Let R_0 be the subring of R generated by $1, a_0, \cdots, a_k$; we have just seen that rt and s are in $R_0[t]$. Since the a_i commute with each other R_0 is a commutative ring. But now from $(1 + rt)(1 + s) = 1$ we know that $1 + rt$ is invertible in $R_0[t]$; invoking Lemma 6.1.2 we get that $a_0, \cdots, a_k$

are all nilpotent. If U denotes

$$\{a \in R \mid at^{n_0} + a_1' t^{n_1} + \cdots + a_k' t^{n_k} \in J$$
$$\text{for some } a_1', \cdots, a_k' \in R\}$$

then U is an ideal of R, contains a_0, so is not (0) and, from what we have just seen, is nil. This proves the theorem.

The method of proof used above can be refined to prove a sharper result, namely: $J(R[t]) = M[t]$ for some appropriate nil ideal M of R.

2. Standard identities. In what follows R will be assumed to be an algebra over a field F. All of it could be developed and generalized in a much broader context but to avoid trifling, annoying and inessential hypotheses we have chosen to work in this narrower framework.

DEFINITION. *An algebra A over a field F is said to satisfy a polynomial identity if there is an $f \neq 0$ in $F[x_1, \cdots, x_d]$, the free algebra over F in the noncommuting variables $x_1, \cdots, x_d$ for some d, such that $f(a_1, \cdots, a_d) = 0$ for all $a_1, \cdots, a_d$ in A.*

We shall then refer to A as a P.I. algebra. Some immediate examples of P.I. algebras are:

1. Any commutative algebra A over F is a P.I. algebra for it satisfies $f(x_1, x_2) = x_1 x_2 - x_2 x_1$.

2. F_2 satisfies

$$f(x_1, x_2, x_3) = (x_1 x_2 - x_2 x_1)^2 x_3 - x_3 (x_1 x_2 - x_2 x_1)^2.$$

P.I. algebras exist in abundance and we shall soon see many general examples of such. We now show that there is no universal identity which holds for all matrix algebras.

LEMMA 6.2.1. *If d is a positive integer and $f \neq 0$ is in*

$F[x_1, \cdots, x_d]$ *then there exists an integer* n *such that* F_n *does not satisfy* f.

Proof. Let f be of degree k and let M be the ideal of $F[x_1, \cdots, x_d]$ generated by all the monomials in $x_1, \cdots, x_d$ of degree larger than k. Hence $A = (F[x_1, \cdots, x_d])/M$ is a finite dimensional algebra over F. Using the regular representation of A we represent it as a subalgebra of F_n for $n = \dim {}_F A$. Since $f \notin M$ its image $\bar{f}$ in A is not 0. Thus there are matrices $a_1, \cdots, a_d$ in F_n such that $f(a_1, \cdots, a_d) \neq 0$.

We generalize the notion of commutativity; in doing so we introduce a special but extremely significant class of polynomial identities.

DEFINITION. *In* $F[x_1, \cdots, x_n]$ *the standard identity of degree* n *is*

$$[x_1, \cdots, x_n] = \sum_{\sigma \in S_n} (-1)^\sigma x_{\sigma(1)} \cdots x_{\sigma(n)}$$

where σ runs over S_n, the symmetric group of degree n, and where $(-1)^\sigma$ is 1 or -1 according as σ is an even or odd permutation.

A is said to satisfy a standard identity if $[x_1, \cdots, x_n]$ vanishes on A for some n. Note that the standard identity of degree 2 is merely $x_1 x_2 - x_2 x_1$; in this sense algebras satisfying a standard identity generalize the class of commutative algebras. We provide ourselves with a ready source of P.I. algebras via

LEMMA 6.2.2. *If* A *is an* n-*dimensional algebra over* F *then* A *satisfies* $[x_1, \cdots, x_{n+1}]$.

Proof. From the very definition of the standard identity we see that it is multilinear in its variables and that it vanishes if two of its arguments are equal. If $u_1, \cdots, u_n$ is a basis of A over F and if $a_1, \cdots, a_{n+1}$ are in

A then we can express each a_i as a linear combination of $u_1, \cdots, u_n$ over F. By the multilinearity of $[x_1, \cdots, x_{n+1}]$ we see that $[a_1, \cdots, a_{n+1}]$ is a linear combination of terms of the form $[u_{i_1}, \cdots, u_{i_{n+1}}]$ where u_{i_j} is chosen from $u_1, \cdots, u_n$. But then in $[u_{i_1}, \cdots, u_{i_{n+1}}]$ two arguments are equal so it vanishes. Hence $[a_1, \cdots, a_{n+1}] = 0$.

COROLLARY 1. F_n satisfies $[x_1, \cdots, x_{n^2+1}]$.

As a point in fact the result in Corollary 1 is much too crude, for Amitsur and Levitzki have shown that F_n actually satisfies $[x_1, \cdots, x_{2n}]$. In the work we shall do here we do not need the sharper estimate.

COROLLARY 2. If A is a commutative algebra over F then A_n satisfies $[x_1, \cdots, x_{n^2+1}]$.

If A is finite dimensional over F, of dimension n, then every element in A satisfies a polynomial of degree $n+1$ over F (n if A should have a unit element). We use this to define the notion of an algebraic algebra of bounded degree over F. A is said to be an *algebraic algebra of bounded degree* over F if there exists an integer n such that given $a \in A$ there exists a polynomial $x^n + \alpha_1 x^{n-1} + \cdots + \alpha_n \in F[x]$ satisfied by a. We generalize the result of Lemma 6.2.2 to

LEMMA 6.2.3. *If A is algebraic of bounded degree over F then A is a P.I. algebra.*

Proof. Suppose that every element b of A satisfies some polynomial of the form $x^n + \alpha_1 x^{n-1} + \cdots + \alpha_n$ where the $\alpha_i \in F$ and n is a fixed integer. For any $a \in A$, commuting it with

$$b^n + \alpha_1 b^{n-1} + \cdots + \alpha_n = 0$$

we get (where $[u, v]$ indicates $uv - vu$)

$$[b^n, a] + \alpha_1[b^{n-1}, a] + \cdots + \alpha_{n-1}[b, a] = 0.$$

Commute this with $[b, a]$; on doing so we get

$$[[b^n, a], [b, a]] + \alpha_1[[b^{n-1}, a], [b, a]] + \cdots$$
$$+ \alpha_{n-2}[[b^2, a], [b, a]] = 0.$$

Commute this with $[[b^2, a], [b, a]]$ and so on, n times, to get a nontrivial identity involving higher commutators of the $[b^i, a]$ and none of the coefficients α_i. In this way we exhibit a specific identity satisfied by A.

Whenever A satisfies a polynomial identity we would like to be able to assume that this identity has some decent form. One such simplification in the nature of the identity is afforded us in

LEMMA 6.2.4. *If A satisfies a polynomial identity of degree d then it satisfies a multilinear identity of degree less than or equal to d.*

Proof. Let A satisfy the identity $f(x_1, \cdots, x_n)$ of degree d. Then it also satisfies the identity

$$g(x_1, \cdots, x_n, x_{n+1}) = f(x_1 + x_{n+1}, x_2, \cdots, x_n)$$
$$- f(x_1, \cdots, x_n) - f(x_{n+1}, \cdots, x_n)$$

which is of lower degree in x_1 than is f. Continuing in this manner we arrive at an identity linear in x_1; in the process we have introduced one new variable in each of the linearizations made. Therefore the degree of the new polynomial obtained is at most d. Now go on to x_2 and repeat the procedure. As we run through all the variables we end up with a multilinear identity of degree at most d which is satisfied by A.

The following lemma is immediate from the definitions of tensor product and multilinearity. We leave its proof to the reader.

LEMMA 6.2.5. *If A satisfies a multilinear identity f then for any extension field, K, of F, $A \otimes_F K$ also satisfies f.*

3. A theorem of Kaplansky. To this point what we have said about polynomial identities has been of a formal or combinatorial flavor. We now turn to the algebraic side of the question. The key result—and the one that renders the subject tractable—is a very important and pretty theorem due to Kaplansky (Theorem 6.3.1). However, first we need some specific information about F_n.

LEMMA 6.3.1. *F_n does not satisfy a polynomial identity of degree less than $2n$.*

Proof. If F_n satisfies a polynomial identity of degree less than $2n$ then, by Lemma 6.2.4, it would satisfy a multilinear identity f of degree $d < 2n$. In fact, we may assume that this identity is homogeneous (prove!) hence we may write f as

$$f = x_1 x_2 \cdots x_d + \sum_{\sigma \neq 1 \in S_d} \alpha_\sigma x_{\sigma(1)} \cdots x_{\sigma(d)}$$

where $\alpha_\sigma \in F$ and S_d is the symmetric group of degree d.

We make a particular choice in F_n for $x_1, x_2, \cdots, x_d$ as follows: $x_1 = e_{11}, x_2 = e_{12}, x_3 = e_{22}, x_4 = e_{23}, \cdots$ where the e_{ij} denote the usual matrix units. Since $d < 2n$ we can make such a choice and, moreover, $x_{\sigma(1)} \cdots x_{\sigma(d)} = 0$ if $\sigma \neq 1$ for the choice of x's made. Thus

$$f(e_{11}, e_{12}, e_{22}, \cdots) = e_{11} e_{12} e_{22} e_{23} \cdots \neq 0,$$

whence F_n does not satisfy f.

We are now able to establish

THEOREM 6.3.1 (Kaplansky). *If A is a primitive algebra satisfying a polynomial identity of degree d then A is a finite dimensional simple algebra over its center, of dimen-*

sion at most $[d/2]^2$ *where* $[d/2]$ *is the largest integer in* $d/2$.

Proof. Since A is primitive then, by Theorem 2.1.4, either it is isomorphic to Δ_n for some integer n and division ring Δ or else Δ_m is a homomorphic image of some subalgebra of A for every integer m. However, subalgebras and homomorphic images of A satisfy the identity of A. Thus in the second possibility above Δ_m would satisfy the identity of A for every m, and hence so would Z_m, where Z is the center of Δ. By Lemma 6.2.1 this is impossible. Hence the first possibility is the only one open to us and so $A \approx \Delta_n$.

Let K be a maximal subfield of Δ. By Theorem 4.2.1 $\Delta \otimes_Z K$ is a dense ring of linear transformations on a vector space over K, hence is a dense ring of K-linear transformations. As we noted earlier, we may suppose that A satisfies a multilinear identity f of degree d or less, hence $A \otimes_Z K$ also satisfies f. The argument given above then shows that $A \otimes_Z K \approx K_n$, since K is its commuting ring. But $\dim_K(A \otimes_Z K) = \dim_Z(A)$ hence $\dim_Z(A) = n^2$. By Lemma 6.3.1, since K_n satisfies f of degree d, $n \leq [d/2]$ hence $\dim_Z(A) \leq [d/2]^2$. Since A is primitive and finite dimensional over Z it is simple. The theorem is now established.

This theorem has many applications and is central in almost all work done on P.I. algebras. We indicate a simple application. Suppose that A is a primitive algebra over F algebraic of bounded degree n. By Lemma 6.2.3, A satisfies a polynomial identity, hence by Theorem 6.3.1 A is a simple algebra finite dimensional over its center. Therefore $A \approx \Delta_m$ for some m. This forces Δ_m to be algebraic of bounded degree n over F and hence over Z. But in Z_m there is always a matrix whose minimal polynomial over Z is of degree m. Consequently

$m \leqq n$. As we did in the proof we can split Δ by tensoring with a maximal subfield and arrive at dim $_Z(A) \leqq n^2$. In particular, if every element in A satisfies a quadratic equation over Z then A is at most 4-dimensional over Z.

If B is a commutative ring then we saw that B_n satisfies a polynomial identity. Is the converse true, that is, if R satisfies a polynomial identity, is $R \subset B_n$ for some commutative ring n? In this generality the answer is no. Let A be an infinite dimensional algebra over a field of characteristic 0 with a basis $u_1, \cdots, u_n, \cdots$ such that $u_i u_j = -u_j u_i$ for all i, j. (This is merely an exterior algebra.) It is easy to show that A satisfies the identity $[[x, y], z]$. However it is also easy to show that A does not satisfy a standard identity. Hence A cannot be contained in B_m for any m and any commutative B. However, if the ring is properly conditioned we do indeed get some imbedding theorem. This is

THEOREM 6.3.2. *Let A be a P.I. algebra having no nonzero nil ideals. Then $A \subset B_m$ where the commutative ring B is a direct sum of fields.*

Proof. Because A has no nonzero nil ideals invoking Theorem 6.1.1 we get that $A[t]$, the polynomial ring in t over A, is semi-simple. Since we may assume that the identity of A is multilinear and since $A[t] = A \otimes_F F[t]$ then $A[t]$ is also a P.I. algebra. Since $A \subset A[t]$ if we could show that $A[t] \subset B_m$ where B is of the required form we would certainly have the result for A. In other words we may assume, without loss of generality, that A is semi-simple. As such, A is a subdirect sum of primitive rings A_α each of which, as a homomorphic image of A, satisfies the same identity f as does A. If f is of degree d then by Kaplansky's theorem each A_α is of dimension at most $[d/2]^2$ over Z_α, its center. As A_α is a simple algebra Z_α must be a field. Using the regular representation of A_α over Z_α we can imbed A_α in the $k \times k$ matrices over Z_α where $k = \dim {}_Z(A_\alpha) \leqq [d/2]^2$. Thus there is an integer m so that each $A_\alpha \subset (Z_\alpha)_m$. If B is the direct product of the Z_α's then we have shown that $A \subset B_m$.

Note that as an immediate consequence of the theorem we get the

COROLLARY. *If A is a P.I. algebra having no nonzero nil ideals then A satisfies a standard identity.*

We close this section with a result about subalgebras of matrix algebras, a result which indicates that by means of identities we can separate out the subalgebras from the matrix algebra itself. But first we need a preliminary result. Let $r(n)$ be the degree of the minimal standard identity satisfied by F_n.

LEMMA 6.3.2. $r(n) \geqq r(n-1) + 2$.

Proof. We imbed F_{n-1} in F_n by sending $a \in F_{n-1}$ into

$$\begin{pmatrix} a & 0 \\ 0 & 0 \end{pmatrix}$$

in F_n. Let $t = r(n-1) - 1$; hence we can find $a_1, \cdots, a_t \in F_{n-1}$ such that $[a_1, \cdots, a_t] \neq 0$. Using the imbedding of F_{n-1} in F_n described above we see that

$$[a_1, \cdots, a_t, e_{kn}, e_{nn}] = [a_1, \cdots, a_t]e_{kn}e_{nn}$$
$$= [a_1, \cdots, a_t]e_{kn}.$$

Since $[a_1, \cdots, a_t]$ is not 0 and of the form

$$\begin{pmatrix} a & 0 \\ 0 & 0 \end{pmatrix}$$

we can find a k such that $[a_1, \cdots, a_t]e_{kn} \neq 0$. But then $[a_1, \cdots, a_t, e_{kn}, e_{nn}] \neq 0$ which yields that $r(n) > t + 2 = r(n-1) + 1$.

THEOREM 6.3.3. *Let R be a simple algebra finite dimensional over its center F. Let $A \subset R$ be a subalgebra over F such that any polynomial identity satisfied by A over P, the prime field of F, is also satisfied by R. Then $A = R$.*

Proof. Let $\overline{F}$ be the algebraic closure of F; our hypothesis carries over to the ring $R \otimes_F \overline{F}$. Hence to prove the theorem we may assume without loss of generality that F is algebraically closed. By Wedderburn's theorem $R = F_n$.

If A is semisimple then by Wedderburn's theorem $A \approx F_{n_1} \oplus \cdots \oplus F_{n_k}$ from which we see that A satisfies all the identities over P which are satisfied by F_{n_t} where $n_t = \max (n_1, \cdots, n_k)$. Hence by our assumption F_n satisfies all the identities over P satisfied by F_{n_t}. By Lemma 6.3.2 we get that $n = n_t$ and so $A = F_n = R$ the desired conclusion.

Suppose, then, that A is not semisimple and its radical N is not (0). Now $A/N \approx F_{n_1} \oplus \cdots \oplus F_{n_k}$ where each $n_i < n$ since dim $A/N <$ dim $A \leq$ dim F_n. If $m = \max (n_1, \cdots, n_k)$ then A/N satisfies the identities of F_m hence it satisfies $[x_1, \cdots, x_r]$ where $r = r(m)$. In other words, for all $a_1, \cdots, a_r \in A$, $[a_1, \cdots, a_r] \in N$. Now N is nilpotent, say $N^t = (0)$. Hence A satisfies the identity over P,

$$b_1 [a_1, \cdots, a_r] b_2 [a_1, \cdots, a_r] \cdots b_t [a_1, \cdots, a_r] = 0.$$

Thus, by hypothesis, F_n satisfies the identity

$$h(y_1, \cdots, y_t, x_1, \cdots, x_r)$$
$$= y_1 [x_1, \cdots, x_r] y_2 [x_1, \cdots, x_r] \cdots y_t [x_1, \cdots, x_r].$$

Since F_n is simple and has no nilpotent ideals, the net result of this is that F_n must satisfy $[x_1, \cdots, x_r]$ (for from the identity, $F_n [a_1, \cdots, a_r]$ is a nilpotent right ideal for $a_1, \cdots, a_r \in F_n$). Since $r = r(m)$ and $m < n$ this is a contradiction. In this way the theorem has been proved.

4. The Kurosh Problem for P.I. algebras. The Kurosh Problem, which is the analog for algebras of the Burnside Problem for groups, asks: If A is an algebraic algebra over F, does a finite number of elements of A always generate a finite dimensional subalgebra of A? In this generality we shall see, in the last chapter, that the answer is no. However, and unlike the situation for groups, in the presence of a polynomial identity the answer is yes. In particular for algebraic algebras of bounded degree the Kurosh Problem has an affirmative answer.

DEFINITION. *The algebra A over F is said to be locally finite if the subalgebra generated over F by a finite number of elements of A is always finite dimensional over F.*

In these terms we ask: when is an algebraic algebra locally finite? We try to find methods of reducing the problem to simpler families of algebras.

LEMMA 6.4.1. *If A is an algebra over F and B an ideal of A such that both B and A/B are locally finite then A is locally finite.*

Proof. Let $a_1, \cdots, a_r$ be a finite set of elements in A and let $\bar{a}_1, \cdots, \bar{a}_r$ be their images in $\overline{A} = A/B$. Since A/B is locally finite $\bar{a}_1, \cdots, \bar{a}_r$ generate a finite dimensional subalgebra of $\overline{A}$ over F; this subalgebra can be spanned over F by $\bar{a}_1, \cdots, \bar{a}_r, \bar{a}_{r+1}, \cdots, \bar{a}_m$. Let $a_1, \cdots, a_r$ and $a_{r+1}, \cdots, a_m$ be inverse images of these elements in A. From the relations $\bar{a}_i \bar{a}_j = \sum \alpha_{ijk} \bar{a}_k$, where $\alpha_{ijk} \in F$, we get that $a_i a_j = \sum \alpha_{ijk} a_k + b_{ij}$ where $b_{ij} \in B$. Since B is an ideal of A all the elements $b_{ij}, a_k b_{ij}, b_{ij} a_k, a_k b_{ij} a_t$ are in B hence, by the local finiteness of B they generate a finite dimensional subalgebra M of A. If W is the subalgebra of A generated by $a_1, \cdots, a_m$ then $M \subset W$ and, moreover, is an ideal of W. Now $\overline{W} = W/M$ is spanned by the elements $\overline{a}_1, \cdots, \overline{a}_m$, the images of $a_1, \cdots, a_m$, which satisfy $\overline{a}_i \overline{a}_j = \sum \alpha_{ijk} \overline{a}_k$, hence $\overline{W}$ is a finite dimensional algebra over F. Both M and W/M being finite dimensional, we infer that W is finite dimensional over F. W certainly contains the subalgebra generated by $a_1, \cdots, a_r$ so this subalgebra is finite dimensional over F. The lemma is proved.

An ideal of A is said to be a *locally finite ideal* if as an algebra in its own right it is locally finite.

LEMMA 6.4.2. *If B and C are locally finite ideals of A then $B + C$ is a locally finite ideal of A.*

Proof. By the standard homomorphism theorems,

$$\frac{B + C}{C} \approx \frac{B}{B \cap C}.$$

However, as a homomorphic image of B, which is locally finite, $B/B \cap C$ must be locally finite. Hence $(B+C)/C$ is locally finite. By assumption C is locally finite; applying Lemma 6.4.1 gives us that $B + C$ is locally finite.

If $\{B_\alpha\}$ is a linearly ordered set of locally finite ideals of A then their union is immediately seen to be locally finite. Hence Zorn's lemma can be applied to give us a maximal locally finite ideal, L, of A. If C is any locally finite ideal then, by the preceding lemma, $L + C$ is also locally finite. By the maximality of L we conclude that $L + C = L$ and so $C \subset L$. We have proved

LEMMA 6.4.3. *A contains a unique maximal locally finite ideal $L(A)$ which contains all the locally finite ideals of A.*

We could call $L(A)$ the locally finite radical of A for it exhibits many radical-like properties. One of these is expressed in

LEMMA 6.4.4. $L(A/L(A)) = (0)$.

Proof. If $\overline{C}$ is a locally finite ideal of $\overline{A} = A/L(A)$ and if C is its inverse image in A then by the standard elementary homomorphism theorems $C \supset L(A)$ and $C/L(A) = \overline{C}$. Since both $\overline{C}$ and $L(A)$ are locally finite, on applying Lemma 6.4.1, we obtain that C is locally finite. Hence $C \subset L(A)$ which leads us to $\overline{C} = (0)$.

We need a slightly deeper property of $L(A)$ than that asserted in Lemma 6.4.3. This is

THEOREM 6.4.1. *If C is a locally finite left (or right) ideal of A then $C \subset L(A)$.*

Proof. In view of Lemma 6.4.4 by passing to $A/L(A)$ we may assume that $L(A) = (0)$. In this reduction our objective then becomes to prove that a locally finite left ideal C must be (0).

Now CA is a two-sided ideal of A; we claim that it is locally finite. Let $x_1, \cdots, x_m$ be in CA; then each $x_i = \sum c_{ij} r_{ij}$ where the $c_{ij} \in C$ and $r_{ij} \in A$. Let $y_{ijmq} = r_{ij} c_{mq}$; these are elements of C hence, from the local finiteness of C, the elements c_{st}, y_{ijmq} generate a finite dimensional algebra W over F. Since $x_i = \sum c_{ij} r_{ij}$ and $x_k = \sum c_{kj} r_{kj}$ we get that

$$x_i x_k = \sum c_{ij} r_{ij} c_{km} r_{km} = \sum c_{ij} y_{ijkm} r_{km} \subset W r_{km}.$$

Hence the product of any two x_i's is contained in $T = \sum W r_{km}$ which is a finite dimensional vector space over F. Now

$$W r_{km} x_t = (W r_{km})(\sum c_{tj} r_{tj}) = \sum W r_{km} c_{tj} r_{tj}$$
$$\subset \sum W y_{kmtj} r_{tj} \subset \sum W r_{tj} \subset T.$$

Thus the subalgebra generated by the x_i's is in T which is finite dimensional over F. In other words we have shown that CA is a locally finite ideal of A. Since $L(A) = (0)$ this yields that $CA = (0)$ hence C is an ideal of A. Being locally finite we conclude that $C = (0)$. This is the theorem.

We are now in a position to start the final steps needed for settling the Kurosh problem for P.I. algebras. The basic and most difficult one is the result of

THEOREM 6.4.2. *Let A be an algebraic algebra over F, finitely generated and satisfying a polynomial identity. If A has no nilpotent elements then $L(A) \neq (0)$.*

Proof. It is an easy exercise to show that the radical of an algebraic algebra is nil. In the presence of the additional hypothesis that A have no nilpotent elements we conclude that A must be semi-simple.

Let $a \neq 0$ be a noninvertible element of A. Then a satisfies some minimal polynomial relation of the form $a^n + \alpha_1 a^{n-1} + \cdots + \alpha_k a^{n-k} = 0$ where $\alpha_k \neq 0$ and where $n - k \geq 1$. But then we see that

$$((a^k + \alpha_1 a^{k-1} + \cdots + \alpha_k)a)^{n-k} = 0;$$

by our hypothesis we end up with

$$a^{k+1} + \alpha_1 a^k + \cdots + \alpha_k a = 0$$

where $\alpha_k \neq 0$. Thus we see that $a = a^2 p(a)$ for some polynomial p over F. The element $e = ap(a)$ is therefore an idempotent and, since $a \neq 0$ and is not invertible, $e \neq 0, 1$. We have produced an idempotent $e \neq 0, 1$ such that $ae = a$.

For rings with no nilpotent elements all idempotents are in the center (see discussion after the proof of Theorem 3.2.2) hence $e \in Z(A)$, the center of A.

Let $M \neq (0)$ be a nonzero ideal of A and let $a_1, \cdots,$

$a_n \in M$. We claim that there is an idempotent e in M such that $a_i e = a_i$ for $i = 1, 2, \cdots, n$. Going by induction on n we suppose that we already have an idempotent $e_1 \in M$ such that $a_1 e_1 = a_1, \cdots, a_{n-1} e_1 = a_{n-1}$. If $a_n e_1 = a_n$ we are done. If not by the argument given above there is an idempotent $e_2 \in M$ such that $(a_n - a_n e_1) e_2 = a_n - a_n e_1$, which is to say, $a_n = a_n(e_1 + e_2 - e_1 e_2)$. Let $e = e_1 + e_2 - e_1 e_2$; the fact that e_1, e_2 are idempotents and so in $Z(A)$ gives the conclusion that e is an idempotent. Also $a_n e = a_n$. Furthermore, for $i < n$ we have

$$a_i e = a_i(e_1 + e_2 - e_1 e_2) = a_i e_1 + a_i e_2 - a_i e_1 e_2 = a_i$$

since $a_i e_1 = a_i$. We have produced the requisite idempotent e in M.

Let P be an ideal of A such that A/P is primitive. Since A/P satisfies a polynomial identity, by Kaplansky's theorem it is a simple algebra finite dimensional over its center. Since A is finitely generated over F so is A/P. Since A/P is algebraic over F, its center, which is a field, it is algebraic over F. These all combine to tell us that A/P is finite-dimensional over F.

Let $\bar{y}_1, \cdots, \bar{y}_m$ be a basis of A/P over F and let $x_1, \cdots, x_n$ generate A. Let $y_1, \cdots, y_m$ in A be inverse images of $\bar{y}_1, \cdots, \bar{y}_m$. We therefore have:

$$(1) \qquad x_i = \sum_{j=1}^{m} \alpha_{ij} y_j + u_i \quad \text{with } \alpha_{ij} \in F, u_i \in F$$

$$(2) \qquad y_i y_j = \sum_{k=1}^{m} \beta_{ijk} y_k + u_{ij} \quad \text{with } \beta_{ijk} \in F, u_{ij} \in P.$$

Let P_0 be the ideal of A generated by all the u_i and u_{ij}; certainly $P_0 \subset P$. Since $x_1, \cdots, x_n$ generate A, by (1) and (2) we get that every element in A has the form $\sum \gamma_i y_i + t$ where the $\gamma_i \in F$ and $t \in P_0$. In particular, if $a \in P$ then $a = \sum \gamma_i y_i + t$ hence $a - t = \sum \gamma_i y_i$ is in P. In A/P this yields $\sum \gamma_i \bar{y}_i = 0$; the $\bar{y}_i$ being independent over F leads us to $\gamma_i = 0$ and so $a = t$. In other words we have shown that $P_0 = P$.

By what we have established earlier in this proof there is an idempotent e in P such that $u_i e = u_i$ and $u_{ij} e = u_{ij}$ for all i, j. Since e is in the center and the u_i, u_{ij} generate $P_0 = P$ as an ideal we get that $pe = p$ for all $p \in P$. Hence $P = Ae$. Since $e \in Z(A)$ the Peirce decomposition $A = Ae \oplus A(1-e)$ is in two-sided ideals; moreover $A/P = A/Ae \approx A(1-e)$. But A/P is finite-dimensional over F, hence $A(1-e)$ is finite-dimensional over F. In consequence $A(1-e) \subset L(A)$ and so $L(A) \neq (0)$.

Everything is now in order for us to give an affirmative answer to the Kurosh problem for P.I. algebras.

THEOREM 6.4.3. *If A is an algebraic algebra over F satisfying a polynomial identity then A is locally finite.*

Proof. Since A satisfies a polynomial identity over F we may assume that it satisfies a multilinear identity of degree d. Our proof will proceed by induction on d. If $d = 2$ it is easy to show that either A is commutative or $A^3 = (0)$ and in either of these cases the theorem is true.

By going to $A/L(A)$ we may assume that $L(A) = (0)$. Since we are only concerned with the finitely generated subalgebras of A we may further assume that A is finitely generated. In short we may reduce to the situation in which A is a finitely generated algebraic algebra over F satisfying a multilinear identity of degree d and for which $L(A) = (0)$. We will show that this is impossible.

If A fails to have nilpotent elements, by the preceding theorem $L(A) \neq (0)$. Therefore we must have nonzero nilpotent elements in A. Let $a \neq 0$ be in A such that $a^2 = 0$ and let T be the left ideal of A generated by a. Thus $Ta = (0)$. Now A satisfies

$$f(x_1, \cdots, x_d) = x_1 q(x_2, \cdots, x_d) + h(x_1, \cdots, x_d)$$

where in $h(x_1, \cdots, x_d)$ the variable x_1 is never the first term of any monomial in the x's.

Let $x_1 = a$, $x_2 = t_2, \cdots, x_d = t_d$ where $t_2, \cdots, t_d \in T$. On substituting these in $f(x_1, \cdots, x_d)$ and using $Ta = (0)$ we arrive at $aq(t_2, \cdots, t_d) = 0$. Let $W = \{x \in T \mid ax = 0\}$; since $TW = (0)$ we have that W is a two-sided ideal of T. Moreover, $\overline{T} = T/W$ satisfies the identity of $q(x_2, \cdots, x_d)$ of degree $d-1$. By our induction, $\overline{T}$ is locally finite. Since $W^2 = (0)$ W is certainly locally finite. From $\overline{T} = T/W$ and W both locally finite we deduce that T is locally finite. Hence $T \neq (0)$ is a locally finite left ideal of A. By Theorem 6.4.1 we conclude that $T \subset L(A)$ whence $L(A) \neq (0)$. This contradicts our hypothesis that $L(A) = (0)$ and proves the theorem.

Although the next result is a mere corollary to Theorem 6.4.3, as the algebra analog of the bounded Burnside problem it has an independent interest; hence we single it out as

THEOREM 6.4.4. *If A is an algebraic algebra of bounded degree over F then it is locally finite.*

Proof. By Lemma 6.2.3. A satisfies a polynomial identity hence the result follows from Theorem 6.4.3.

References

1. S. A. Amitsur and J. Levitzki, Minimal identities for algebras, *Proc. Amer. Math. Soc.*, 1 (1950) 449–463.

2. Nathan Jacobson, *Structure of Rings*, Amer. Math. Soc., Colloq. Publ., 37 (1964).

3. ————, Structure theory for algebraic algebras of bounded degree, *Annals of Math.*, 46 (1945) 695–707.

4. Irving Kaplansky, Rings with polynomial identity, *Bull. Amer. Math. Soc.*, 54 (1948) 575–580.

5. ————, Groups with representations of bounded degree, *Canadian J. Math.*, 1 (1949) 105–112.

6. J. Levitzki, A problem of A. Kurosh, *Bull. Amer. Math. Soc.*, 52 (1946) 1033–1035.

CHAPTER 7

GOLDIE'S THEOREM

Goldie has recently proved several theorems which give penetrating information about the nature of rings subject to certain chain conditions. In the theory of rings with ascending chain conditions these theorems assume the role played by the Wedderburn theorems in the theory of rings with descending chain conditions. In fact these results are extensions of the Wedderburn theorems to a wider class of rings. Procesi gave a short and highly conceptual proof of the first of Goldie's results; we then showed how one can easily obtain the second from the first.

The proof of these theorems that we shall present here is due to Procesi and Small. It is simple, short and neat, yet it is not nearly as revealing as the proof given earlier by Procesi. For one thing, that proof actually also gives us Wedderburn's theorem for simple rings. For another, the techniques developed there allow us to proceed to a variety of results; to cite an instance one readily obtains a recent result of Faith and Utumi. Be that as it may since our primary interest will be in the Goldie theorems we shall content ourselves here with the Procesi-Small approach to the subject.

1. Ore's Theorem. As is elementary and well known, one can imbed a commutative integral domain in a field, the field being nothing else than the fractions created from the elements of the domain. Ore gave the appropriate conditions in order that this be possible for noncommutative rings without zero divisors. We shall give an account of this—rather, of a more general situation— below. But first a few definitions.

DEFINITION. *An element in a ring R is said to be regular if it is neither a left nor right zero divisor in R.*

DEFINITION. *A ring $Q(R) \supset R$ is said to be a left quotient ring for R if:*
1. *every regular element in R is invertible in $Q(R)$.*
2. *every $x \in Q(R)$ is of the form $x = a^{-1}b$ where $a, b \in R$ and a is regular.*

If $Q(R)$ is a left quotient ring of R we say that R is a *left order* in $Q(R)$. We now prove Ore's theorem.

THEOREM 7.1.1. *A necessary and sufficient condition that R have a left quotient ring is: given a, $b \in R$ with b regular then there exist a_1, $b_1 \in R$ with b_1 regular such that $b_1 a = a_1 b$.*

Proof. The condition of the theorem is usually called the *(left) Ore condition.* Now to the proof. If $Q(R)$ exists then for b regular in R the element ab^{-1} is in $Q(R)$ hence $ab^{-1} = b_1^{-1}a_1$ for a_1, $b_1 \in R$ with b_1 regular; transposing we get $b_1 a = a_1 b$, in short, the Ore condition holds in R.

Now for the other, and more interesting, direction. Suppose that the Ore condition holds in R. Let $\mathfrak{M} = \{(a, b) \mid a, b \in R, b \text{ regular}\}$. In $\mathfrak{M}$ we define a relation $(a, b) \sim (c, d)$ if $d_1 a = b_1 c$ where $b_1 d = d_1 b$, b_1 regular. Hence it also follows that d_1 is regular. We claim that this is independent of the particular b_1, d_1 which give the common left multiple of d and b. For if $b_2 d = d_2 b$ pick e_1, e_2 regular so that $e_2 b_2 = e_1 b_1$ then $e_2 d_2 b = e_2 b_2 d = e_1 b_1 d = e_1 d_1 b$; since b is regular we end up with $e_2 d_2 = e_1 d_1$. From $d_1 a = b_1 c$ we get $e_2 e_2 a = e_1 d_1 a = e_1 b_1 c = e_2 b_2 c$; the regularity of e_2 permits us to conclude that $d_2 a = b_2 c$ as claimed.

It is easy to verify that the relation we have defined on $\mathfrak{M}$ is an equivalence relation. Let the class of (a, b) be written as a/b. We let M be the set of equivalence classes in $\mathfrak{M}$. In M we introduce operations to render it a ring.

For a/b, c/d in M define $a/b + c/d = (d_1 a + b_1 c)/(b_1 d)$ where $d_1 b = b_1 d$, both b_1, d_1 regular. We similarly define $(a/b)(c/d) = (a_1 c)/(g_1 b)$ where $g_1 a = a_1 d$, g_1 regular in R.

We leave it to the reader to verify that these operations are well defined and that M satisfies all the requisite properties of $Q(R)$ spelled out in the definition of $Q(R)$.

2. Goldie's theorems. In this section we shall give the Procesi-Small proof of the Goldie theorems. This proof uses a bare minimum of technique and strips these important results down to their essentials.

We begin with some notation and definitions.

In any ring R, for a nonempty subset S of R, let $l(s) = \{x \in R \mid xs = 0$ for all $s \in S\}$. We call $l(s)$ the *left annihilator* of S and term a left ideal λ of R a left annihilator if $\lambda = l(S)$ for some appropriate S in R. We similarly define the *right annihilator* $r(S)$ of S and speak of a right ideal as a right annihilator.

DEFINITION. *A ring R is said to be a (left) Goldie ring if:*

1. *R satisfies the ascending chain condition on left annihilators.*
2. *R contains no infinite direct sums of left ideals.*

Clearly a left Noetherian ring—that is, one satisfying the ascending chain condition on left ideals—is a Goldie ring. The converse is not true.

DEFINITION. *A left ideal I of R is said to be essential if I intersects every nonzero left ideal of R in a nontrivial fashion.*

Some people prefer to describe such ideals as *large*

rather than as essential but we shall stick with the terminology used above.

Finally, a ring R is said to be *semiprime* if it has no nonzero nilpotent ideals.

Before continuing, one remark is in order; the ascending chain condition on left annihilators is equivalent to the descending chain condition on right annihilators. This remark, although trivial, is important in what follows.

We begin with the key result in this approach to Goldie's work,

LEMMA 7.2.1. *Let R be a semiprime ring satisfying the ascending chain condition on left annihilators. If $A \supset B$ are left ideals of R and $r(A) \neq r(B)$ then there exists an $a \in A$ such that $Aa \neq (0)$ and $Aa \cap B = (0)$.*

Proof. As we pointed out above, since R satisfies the ascending chain condition on left annihilators it satisfies the descending chain condition on right annihilators Since $A \supset B$ we have $r(A) \subset r(B)$ and by hypothesis this containing relation is proper.

Let U be a right annihilator minimal with respect to being contained in $r(B)$ and properly containing $r(A)$. By its choice we have $AU \neq (0)$; since R has no nilpotent ideals $AUAU \neq (0)$. Pick $ua \in UA$ so that $Aua U \neq (0)$. We claim that $Aua \cap B = (0)$. If not there is an $x \in A$ such that $xua \neq 0$ is in $Aua \cap B$. Since $x \in A$, $r(x) \supset r(A)$; consider $r(x) \cap U$. As the intersection of two right annihilators it is a right annihilator; it contains $r(A)$, is contained in U and, moreover, since $uaU \subset r(x)$ but $uaU \not\subset r(A)$ it properly contains $r(A)$. From the minimality of U we deduce that $r(x) \cap U = U$ hence $U \subset r(x)$. This says that $xU = (0)$ which clearly contradicts $xua \neq 0$. The lemma is proved.

The lemma has two important corollaries.

COROLLARY 1. *Let R be as in the lemma; if Rx and Ry are essential left ideals then Rxy is essential.*

Proof. Let $A \neq (0)$ be a left ideal of R and let $\overline{A} = \{r \in R \mid ry \in A\}$; since Ry is essential $\overline{A} \neq (0)$ and $\overline{A}y = Ry \cap A \neq (0)$. Clearly, from its definition $\overline{A} \supset l(y)$; now $\overline{A}y \neq (0)$ and $l(y)y = (0)$ hence we can apply Lemma 7.2.1 to obtain a left ideal $T \neq (0) \subset \overline{A}$ such that $T \cap l(y) = (0)$. Let $\overline{T} = \{r \in R \mid rx \in T\}$; the essentiality of Rx implies that $\overline{T}x = Rx \cap T \neq (0)$. Thus $\overline{T}xy \neq (0)$; being in A we have shown that $Rxy \cap A \neq (0)$ and so Rxy is essential.

COROLLARY 2. *Let R be as in the lemma. Then if Ra is essential in R a must be regular.*

Proof. That $r(a) = (0)$ follows from the special case in which $A = R$, $B = Ra$ and the essentiality of Ra.

We now consider $l(a)$. By the ascending chain condition on left annihilators there is an integer n such that $l(a^n) = l(a^{n+1})$. If $x \in Ra^n \cap l(a)$ then $x = ya^n$ and $o = xa = ya^{n+1}$ placing $y \in l(a^{n+1}) = l(a^n)$, that is $ya^n = 0$ and so $x = ya^n = 0$. By the first corollary above Ra^n is essential; the net result from $Ra^n \cap l(a) = 0$ then must be that $l(a) = (0)$.

For the remainder of this section R will be a semiprime left Goldie ring.

Lemma 7.2.2. *R satisfies the descending chain condition on left annihilators.*

Proof. Let $L_1 \supset L_2 \supset \cdots \supset L_n \supset \cdots$ be a properly descending chain of left annihilators; hence $r(L_i) \neq r(L_{i+1})$. Applying Lemma 7.2.1 there exist left ideals of R, $(0) \neq C_n \subset L_n$, such that $C_n \cap L_{n+1} = (0)$. The C_n thus form a direct sum of left ideals; since R is a Goldie ring, this sum cannot be infinite hence the chain of annihilators must terminate.

In a corollary above we saw that essentiality of a principal left ideal implied the regularity of its generating element. We now turn the affair around and show that regular elements generate essential left ideals.

LEMMA 7.2.3. *If $l(c) = (0)$ then Rc is essential and so c is regular.*

Proof. Let $A \neq (0)$ be a left ideal of R satisfying $A \cap Rc = (0)$. Since $l(c) = (0)$ we claim that the Ac^n form a direct sum; for if $a_0 + a_1c + \cdots + a_nc^n = 0$ with the $a_i \in A$ then $a_0 \in A \cap Rc = (0)$ from which $a_1c + \cdots + a_nc^n = 0$. Since $l(c) = (0)$ this yields $a_1 + a_2c + \cdots + a_nc^{n-1} = 0$; repeat the argument to get that each $a_i = 0$. The absence of infinite direct sums forces the conclusion that $Rc \cap A \neq (0)$ and so Rc is essential. By Corollary 2 to Lemma 7.2.1, c must be regular.

A two-sided ideal S of R is said to be an *annihilator ideal* if S is the left annihilator of some left ideal T. From $ST = (0)$ we have $(TS)^2 = TSTS = (0)$ and so, since R is semiprime, $TS = (0)$. Let us recall that a ring is a prime ring if the product of two nonzero ideals is nonzero.

We now try to reduce from a semiprime ring to certain well behaved prime subrings.

LEMMA 7.2.4. *A nonzero minimal annihilator ideal of R is a prime Goldie ring; moreover there is a finite direct sum of such ideals which is an essential left ideal of R.*

Proof. Let $(0) \neq S$ be a minimal annihilator ideal. If $T \neq (0)$ is a left ideal of S then, since R is semiprime, $ST \neq (0)$; but $ST \subset T$ is a left ideal of R. In short, we immediately have that S has no infinite direct sums of left ideals. Since subrings of R inherit the ascending chain condition on left annihilators we get that S is a Goldie ring.

Suppose that A, B are ideals of S such that $AB = (0)$.

Thus, from $SB \subset B$, we have $ASB = (0)$ and so $A \subset l(SB) \cap S$. But $l(SB) \cap S$ is an annihilator ideal and so, if $A \neq (0)$, by the minimality of S we conclude that $l(SB) \supset S$. This implies that $(SB)^2 = (0)$ and so $SB = (0)$ ending up with $B = (0)$. In other words, S is a prime Goldie ring.

Let $A = S_1 \oplus \cdots \oplus S_n$ be a maximal direct sum of minimal annihilator ideals; we claim that A is essential. For if $A \cap K = (0)$ where $K \neq (0)$ is a left ideal of R then $AK \subset A \cap K = (0)$ hence $K \subset r(A)$. Since R is semiprime, $A \cap r(A) = 0$ and so in $r(A)$ we can find a nonzero minimal annihilator ideal which fails to intersect A. But in that case we can lengthen the direct sum giving A, contradicting the choice of A.

In every approach to Goldie's theorems one eventually gets down to a close interrelation between regularity and essential left ideals. We, too, are now at that point. In truth it is the crux of the theorem. We prove

LEMMA 7.2.5. *If I is an essential left ideal of R then it contains a regular element.*

Proof. We first dispose of the special case in which R is a prime ring. Pick $a \in I$ so that $l(a)$ is minimal; we may do this because of Lemma 7.2.2. If a is regular we are done; if not, by Corollary 2 to Lemma 7.2.2 $Ra \cap J = (0)$ for some left ideal $J \neq (0)$ of R. Since I is essential, $I \cap J \neq (0)$ hence we may suppose that $J \subset I$. If $x \in J$ and if $b \in l(a+x)$ then $b(a+x) = 0$ hence $ba = -bx \in Ra \cap J = (0)$, which is to say that $b \in l(a) \cap l(x)$. By the minimality of $l(a)$ we get that $l(x) \supset l(a) \cap l(x) \supset l(a)$ and so $l(a)x = (0)$ for all $x \in J$. But in a prime ring the fact that $l(a)J = (0)$ and $J \neq (0)$ is a left ideal, forces $l(a) = (0)$. Lemma 7.2.3 assures us that a is regular. This settles the result for prime rings.

If R is semiprime let $A = S_1 \oplus \cdots \oplus S_n$ be as in

Lemma 7.2.4. Since S_i is prime and since $I \cap S_i$ is clearly essential in S_i, by the argument above $I \cap S_i$ contains an element r_i regular in S_i. We claim that $r = r_1 + \cdots + r_n$ is regular in R for if $l(r) \neq (0)$, by the essentiality of A, $l(r) \cap A \neq (0)$. Hence there is an element $0 \neq t = t_1 + \cdots + t_n$ in A with $tr = 0$. But $tr = t_1 r_1 + \cdots + t_n r_n$; the directness of $S_1 \oplus \cdots \oplus S_n$ yields that each $t_i r_i = 0$. Since r_i is regular in S_i this in turn yields that $t_i = 0$ and so $t = 0$. With this contradiction the lemma is established.

We are ready to settle the theorems of Goldie.

THEOREM 7.2.1. *Let R be a semiprime left Goldie ring; then R has a left quotient ring $Q = Q(R)$.*

Proof. Suppose that $a \in R$ is regular and $b \in R$. By Lemma 7.2.3, Ra is essential. If $M = \{r \in R \mid rb \in Ra\}$ it is immediate that M, too, is essential. Hence by Lemma 7.2.3, M contains a regular element c. From the definitions of M, $cb = da$; we have shown that R satisfies the Ore condition. By Theorem 7.1.1 the ring of left quotients, $Q(R)$ exists.

We note two ready facts about Q:

1. If I is a left ideal of Q then $I = Q(I \cap R)$.

2. If $A_1 \oplus \cdots \oplus A_n$ is a direct sum of left ideals in R then $QA_1 + \cdots + QA_n$ is a direct sum of left ideals in Q. (To show this prove: given $x_1, \cdots, x_k \in Q$ then we can find $a \in R$ regular, $b_1, \cdots, b_k \in R$ such that $x_i = a^{-1} b_i$.)

We now are able to determine the explicit structure of Q. This is the important and beautiful theorem due to Goldie,

THEOREM 7.2.2 (Goldie). *Q is a semisimple ring satisfying the descending chain condition on left ideals.*

Proof. Let $I \neq (0)$ be a left ideal of Q; then there is a left ideal K in R such that $(I \cap R) \oplus K$ is essential in R. (For K we can take the complement of $I \cap R$ in the longest direct sum containing $I \cap R$ as a constituent.) Hence $(I \cap R) \oplus K$, by Lemma 7.2.5, has a regular element; this forces $Q = Q(I \cap R) \oplus K = I \oplus QK$. If 1 is the unit element of Q then $1 = i + k$ with $i \in I$, $k \in QK$; as usual, the directness of this sum yields that $i^2 = i \neq 0$, $ik = 0$. Clearly $Qi = I$, that is, I is generated by an idempotent. In particular, every left ideal of Q is principal hence Q certainly must be a left Goldie ring (in fact, a left Noetherian ring). By the fact that every left ideal of Q is generated by an idempotent we have that Q has no nilpotent ideals so is semiprime.

Let $I \neq (0)$ be a left ideal of Q; we have seen that $I = Qe$ where $e^2 = e$. Now $r(Qe) = r(e) = (1 - e)Q$ and $l(r(Qe)) = l((1 - e)Q) = Qe = I$, hence every left ideal of Q is a left annihilator. By Lemma 7.2.2, Q satisfies the descending chain condition on left annihilators, *hence on all left ideals*. We have now shown that Q is indeed a semisimple Artinian ring.

Goldie's theorem has a converse which we now prove.

THEOREM 7.2.3. *Let R be a left order in S, where S is a semisimple Artinian ring. Then R is a semiprime Goldie ring. Moreover, if S is simple then R is prime.*

Proof. We first show that R is a Goldie ring. Since S is a semisimple Artinian ring every left ideal in S is generated by an idempotent; thus S satisfies the ascending chain condition on all left ideals and so, in particular, the ascending chain condition on left annihilators. Since this latter condition is inherited by the subring R, R enjoys the ascending chain condition on left annihilators.

Now let $\lambda_1, \cdots, \lambda_k$ form a direct sum of left ideals in R; we claim that $S\lambda_1, \cdots, S\lambda_k$ form a direct sum of left ideals of S. To see this, if $s_1 a_1 + \cdots + s_k a_k = 0$ with $s_i \in S$, $a_i \in \lambda_i$ then, as we pointed out earlier, there is a d regular in R, $b_i \in R$ such that each $s_i = d^{-1} b_i$; but then we have $b_1 a_1 + \cdots + b_k a_k = 0$. The directness of the sum in R gives that each $b_i a_i = 0$ hence $s_i a_i = d^{-1} b_i a_i = 0$. We have shown that $S\lambda_1, \cdots, S\lambda_k$ form a direct sum; the absence of infinite direct sums in S now implies the absence of infinite direct sums in R. R has been shown to be a Goldie ring.

Let $N \neq (0)$ be a nilpotent ideal in R with, say, $N^m = (0)$, $N^{m-1} \neq (0)$. The ideal SNS of S is not (0) since S has a unit element and so, because S is semisimple and Artinian, $SNS = eS$ where e is a central idempotent in S. Now $e = \sum a_i u_i b_i$ with $u_i \in N$, a_i, $b_i \in S$; we can find a regular in R, $c_i \in R$ such that each $a_i = a^{-1} c_i$ hence

$$e = a^{-1} \sum c_i u_i b_i = a^{-1} \sum w_i b_i$$

where $w_i = c_i u_i \in N$. Since e is central in S, $ae = ea$ and so

$$N^{m-1} ea = N^{m-1} ae = N^{m-1} \left(\sum w_i b_i \right) = (0)$$

since $N^{m-1} w_i \subset N^m = (0)$. But a is regular in R yielding that $N^{m-1} e = (0)$. However

$$(N^{m-1} S)^2 = N^{m-1} S N^{m-1} S \subset N^{m-1} SNS = N^{m-1} eS = (0);$$

since S is semisimple it has no nilpotent right ideals, which forces $N^{m-1} S = (0)$ and so $N^{m-1} = (0)$. This contradiction proves that $N = (0)$. Hence R is semiprime.

What further can one say when S is simple? Let $A \neq (0)$ be an ideal of R; then SAS as a nonzero ideal of S must be S. From $SAS = S$ and the fact that S has a unit element 1 we get $1 = \sum b_i a_i c_i$ with $a_i \in A$, b_i, $c_i \in S$. As we have done several times before, there exists b regular in R such that each $b_i = b^{-1} d_i$ for some $d_i \in R$. If for some ideal B of R we have $BA = (0)$ then $BAS = (0)$, but $1 = b^{-1} \sum d_i a_i c_i$ so that $b = \sum d_i a_i c_i \in AS$. Therefore

Bb being in BAS must be (0). Since b is regular this forces $B = (0)$. We have proved that R is a prime ring. This completes the proof of the converse of Goldie's theorem.

3. Ultra-Products and a Theorem of Posner. Logicians have introduced and successfully used the notion of an ultra-product in their work. Amitsur, in his recent papers, has shown how this notion can usefully be exploited in ring theory. In particular one can get a nice series of imbedding theorems for rings from such an approach. One striking application of this idea is to a theorem due to Posner. This theorem accurately describes the structure of a prime ring satisfying a polynomial identity. In many situations arising in ring theory it can be used even more effectively than the theorem of Kaplansky describing primitive rings with polynomial identities. We follow a line of approach due to Amitsur.

DEFINITION. *Let $A = \{\alpha\}$ be a nonempty set; a set of subsets $\mathfrak{F} = \{S\}$ of A is called a filter if:*
1. *$\emptyset$, the empty set, is not in $\mathfrak{F}$.*
2. *$S_1, S_2 \in \mathfrak{F}$ implies that $S_1 \cap S_2 \in \mathfrak{F}$.*
3. *$S \in \mathfrak{F}$ and $T \supset S$ then $T \in \mathfrak{F}$.*

A filter is called an *ultra-filter* if it is maximal in the natural ordering of filters given by: $\mathfrak{F}_1 \leq \mathfrak{F}_2$ if $S \in \mathfrak{F}_1$ implies that $S \in \mathfrak{F}_2$. It is clear, using a Zorn lemma argument, that any filter can be enlarged to an ultra-filter. Ultra-filters have a simple characterization, namely

LEMMA 7.3.1. *A filter $\mathfrak{F}$ on A is an ultra-filter if and only if given any subset T of A either $T \in \mathfrak{F}$ or its complement, $A - T$, is in $\mathfrak{F}$.*

Proof. If $\mathfrak{F}$ is a filter with the property that for any subset T of A either $T \in \mathfrak{F}$ or $A - T \in \mathfrak{F}$ then $\mathfrak{F}$ is an ultra-filter; for if $\mathfrak{F}'$ is a properly larger filter and $T \in \mathfrak{F}'$, $T \notin \mathfrak{F}$ then $A - T \in \mathfrak{F}$ and so $A - T \in \mathfrak{F}'$. But then $\varnothing = T \cap (A - T) \in \mathfrak{F}'$, which is false from the definition of a filter.

Let $\mathfrak{F}$ be an ultra-filter and suppose that for some subset T, of A, $T \notin \mathfrak{F}$. Let $T' = A - T$. If $T \cap S \neq \varnothing$ for all $S \in \mathfrak{F}$ we can then generate a filter from T and $\mathfrak{F}$ properly larger than $\mathfrak{F}$, contradicting the maximality of $\mathfrak{F}$. So $T \cap S_1 = \varnothing$ for some $S_1 \in \mathfrak{F}$. If $T \notin \mathfrak{F}$ we would have an $S_2 \in \mathfrak{F}$ such that $T \cap S_2 = \varnothing$. But then

$$\phi \neq S_1 \cap S_2 = A \cap S_1 \cap S_2 = (T \cup T') \cap S_1 \cap S_2$$
$$= (T' \cap S_1 \cap S_2) \cup (T \cap S_1 \cap S_2) = \phi,$$

a contradiction. Hence $T' \in \mathfrak{F}$.

Let R_α be a set of rings indexed by $\alpha \in A$ and let $\overline{R} = \pi_{\alpha \in A} R_\alpha$ be the complete product of the R_α's. As usual we consider $\overline{R}$ as the ring of functions $f: A \rightarrow \cup_{\alpha \in A} R_\alpha$ with $f(\alpha) \in R_\alpha$ for all $\alpha \in A$, under pointwise addition and multiplication. Let $\mathfrak{F}$ be a filter on A; for $f, g \in \overline{R}$ we define $f \sim g$ if $\{ \alpha \in A \,|\, f(\alpha) = g(\alpha) \}$ is in the filter. This defines an equivalence relation on $\overline{R}$. Let $I = \{ f \in \overline{R} \,|\, f \sim 0 \}$; I is an ideal of $\overline{R}$. We define $\pi R_\alpha / \mathfrak{F}$ to be $\overline{R}/I$. When $\mathfrak{F}$ is an ultra-filter we call $\pi R_\alpha / \mathfrak{F}$ an ultra-product of the R_α's.

For us the ultra-product assumes an importance for it behaves well in case each R_α is a primitive ring. This is more explicitly expressed in Theorem 7.3.1 to follow. Note first that we can speak about ultra-products, not only of rings, but of vector spaces, groups, modules.

THEOREM 7.3.1. *For* $\alpha \in A$ *let* R_α *be a primitive ring with faithful irreducible module* V_α *and commuting ring* D_α *of* R_α *on* V_α. *Then for any ultra-filter* $\mathfrak{F}$ *on* A, $R = \pi R_\alpha / \mathfrak{F}$ *is a primitive ring with faithful irreducible module* $V = \pi R_\alpha / \mathfrak{F}$ *and commuting ring* $D = \pi D_\alpha / \mathfrak{F}$ *of* R *on* V.

Proof. We shall only prove the primitivity of R leaving the proof to the reader that D is indeed the commuting ring of R on V.

Let $\overline{R} = \pi R_\alpha$ be the complete product of the R_α and $\overline{V} = \pi V_\alpha$ that of the V_α. We let $\overline{R}$ act on $\overline{V}$ by $(vr)(\alpha) = v(\alpha)r(\alpha)$ for $v \in \overline{V}$, $r \in \overline{R}$, $\alpha \in A$. If $r \equiv 0(\mathfrak{F})$ and $v \in \overline{V}$ then $\{\alpha \in A \mid (vr)(\alpha) = 0\}$ contains $\{\alpha \in A \mid r(\alpha) = 0\}$; this second set is in the filter hence $\{\alpha \in A \mid (vr)\alpha = 0\}$ is in $\mathfrak{F}$ and so $vr \equiv 0(\mathfrak{F})$. Similarly, if $v \equiv 0(\mathfrak{F})$ and $r \in \overline{R}$ then $vr \equiv 0(\mathfrak{F})$. We therefore get that $R = \pi R_\alpha / \mathfrak{F}$ acts on $V = \pi V_\alpha / \mathfrak{F}$ and that V is an R-module.

We claim that V is an irreducible R-module. Let $v \not\equiv 0(\mathfrak{F})$ in $\overline{V}$ and let $u \in \overline{V}$. If $T_0 = \{\alpha \in A \mid v(\alpha) \neq 0\}$ then since $v \not\equiv 0(\mathfrak{F})$, $\{\alpha \in A \mid v(\alpha) = 0\}$ is not in $\mathfrak{F}$ hence T_0, its complement, must be in $\mathfrak{F}$. For $\alpha \in T_0$ pick $r_\alpha \in R_\alpha$ so that $v(\alpha)r_\alpha = u(\alpha)$; we can find such an r_α since $v(\alpha) \neq 0$ and R_α acts primitively on V_α, that is, $v(\alpha)R_\alpha = V_\alpha$. If $r \in \overline{R}$ is defined by $r(\alpha) = r_\alpha$ then $\{\alpha \in A \mid (vr - u)(\alpha) = 0\}$ contains T_0 so must be in $\mathfrak{F}$. Thus $vr \equiv u(\mathfrak{F})$. We have shown that R acts transitively on V hence V is an irreducible R-module.

We further assert that R acts faithfully on V for if $vr = 0$ for all $v \in V$ for some $r \in R$ then if r_1 is a representative of r in $\overline{R}$ then $\{\alpha \in A \mid (\bar{v}r_1)(\alpha) = 0 \text{ all } \bar{v} \in \overline{V}\}$ is in $\mathfrak{F}$ and is contained in $\{\alpha \in A \mid r_1(\alpha) = 0\}$ since R_α is faithful on V_α. Hence $\{\alpha \in A \mid r_1(\alpha) = 0\}$ is in $\mathfrak{F}$ leading to $r_1 \equiv 0(\mathfrak{F})$, that is, $r = 0$. The theorem is proved.

We now turn to Posner's theorem.

LEMMA 7.3.2. *If* R *is a prime ring satisfying a polynomial identity over its centroid then* R *is a left and right Goldie ring.*

Proof. From the results in Chapter 6 we know that we may assume that R satisfies a multilinear identity $p(x_1, \cdots, x_n) = \sum_{\sigma \in S_n} \alpha_\sigma x_{\sigma(1)} \cdots x_{\sigma(n)}$ where the α_σ are in the centroid of R and σ runs over the symmetric group S_n.

We assert that no direct sum of left ideals in R can have length n. Let $I_1 \oplus \cdots \oplus I_n$ be such a direct sum of nonzero left ideals of R; pick $x_i \in I_i$. From

$$0 = p(x_1, \cdots, x_n) = \sum \alpha_\sigma x_{\sigma(1)} \cdots x_{\sigma(n)}$$

and from the directness of the sum we get

$$\left(\sum_{\sigma \in S_{n-1}} \alpha_\sigma x_{\sigma(1)} \cdots x_{\sigma(n-1)} \right) x_n = 0$$

for all $x_n \in I_n$. Since R is prime this forces

$$\sum_{\sigma \in S_{n-1}} \alpha_\sigma x_{\sigma(1)} \cdots x_{\sigma(n-1)} = 0$$

$$\text{for } x_i \in I_i, \quad i = 1, \cdots, n-1.$$

Repeat the argument; we eventually get that $I_1 = (0)$, a contradiction. We similarly get no infinite direct sums of right ideals in R.

We now consider the ascending chain condition on left annihilators. Let $I_1 \subset I_2 \subset \cdots \subset I_n$ be a properly ascending chain of left annihilators, where $I_i = l(K_i)$ and $I_i \neq I_{i+1}$.

By assumption $I_i K_{i-1} \neq (0)$ and $K_1 \supset K_2 \supset \cdots \supset K_n$. In $p(x_1, \cdots, x_n)$ put $x_1, \cdots, x_{n-1} \in I_n$, $x_n \in K_n$. Since $I_n K_n = (0)$ the only nonzero contribution to $0 = p(x_1, \cdots, x_n) = \sum \alpha_\sigma x_{\sigma(1)} \cdots x_{\sigma(n)}$ comes from those terms where x_n is in the leftmost position. Hence

$$K_n \left(\sum_{\sigma \in S_{n-1}} \alpha_\sigma x_{\sigma(1)} \cdots x_{\sigma(n-1)} \right) = (0)$$

for $x_1, \cdots, x_{n-1} \in I_n$. Since $K_n \neq (0)$ is a right ideal in a prime ring its right annihilator must be (0). Thus

$$\sum_{\sigma \in S_{n-1}} \alpha_\sigma x_{\sigma(1)} \cdots x_{\sigma(n-1)} = 0$$

for $x_i \in I_n$.

Since R is prime

$$(0) \neq K_{n-1} I_{n-1} \subset I_{n-1} \cap K_{n-1};$$

pick $x_1, \cdots, x_{n-2} \in I_{n-1} \subset I_n$ and $0 \neq x_{n-1} \in K_{n-1} \cap I_{n-1}$. From

$$\sum_{\sigma \in S_{n-1}} \alpha_\sigma x_{\sigma(1)} \cdots x_{\sigma(n-1)} = 0 \quad \text{and} \quad x_i x_{n-1} = 0$$

for $i < n-1$ we get

$$x_{n-1} \sum_{\sigma \in S_{n-1}} \alpha_\sigma x_{\sigma(1)} \cdots x_{\sigma(n-2)} = 0$$

for $x_{n-1} \in K_{n-1} I_{n-1}, x_1, \cdots, x_{n-2} \in I_{n-1}$. Hence

$$K_{n-1} I_{n-1} \left(\sum_{\sigma \in S_{n-2}} \alpha_\sigma x_{\sigma(1)} \cdots x_{\sigma(n-2)} \right) = (0)$$

for $x_1, \cdots, x_{n-2} \in I_{n-1}$; since R is prime we get

$$I_{n-1} \left(\sum_{\sigma \in S_{n-2}} \alpha_\sigma x_{\sigma(1)} \cdots x_{\sigma(n-2)} \right) = (0).$$

In particular if $x_1, \cdots, x_{n-3} \in I_{n-2}$ and $x_{n-2} \in K_{n-2} I_{n-2}$ we get

$$I_{n-1} K_{n-2} I_{n-2} \left(\sum_{\sigma \in S_{n-2}} \alpha_\sigma x_{\sigma(1)} \cdots x_{\sigma(n-3)} \right).$$

Continuing we get $I_{n-1} K_{n-2} I_{n-2} \cdots = (0)$. Since $I_i K_{i-1} \neq (0)$ this gives a contradiction.

We can give a similar argument for right annihilators.

In view of the lemma and Goldie's theorem a prime

ring satisfying a polynomial identity over its centroid is an order in a simple Artinian ring, hence, by Wedderburn's theorem, in D_n for some division ring D. What does the polynomial identity force on D? This is precisely the content of Posner's theorem.

THEOREM 7.3.2. *Suppose that R is a prime ring satisfying a polynomial identity over its centroid. Then R can be imbedded as an order in D_n where D is a division algebra finite-dimensional over its center.*

Before proving this theorem we need a result which has some independent interest.

THEOREM 7.3.3. *Let R be a ring satisfying the ascending chain condition on left annihilators. If R has a nonzero nil ideal then it has a nonzero nilpotent ideal.*

Proof. Let $A \neq (0)$ be a nil ideal of R and pick $a \neq 0$ in A such that $l(a)$ is maximal over all elements in A. If $x \in R$ and $ax \neq 0$ then since $l(ax) \supset l(a)$ and $ax \in A$ we have $l(ax) = l(a)$. Now ax is nilpotent so $(ax)^{k-1} \neq 0$, $(ax)^k = 0$ for some k, hence $ax \in l(ax)^{k-1}) = l(a)$ since $0 \neq (ax)^{k-1} = ay$. In short, $axa = 0$. Thus for any $x \in R$ $axa = 0$; the ideal of R generated by a is therefore nilpotent.

We now proceed with the proof of Posner's theorem.

Let R be a prime ring satisfying a polynomial identity over its centroid. Therefore R certainly has no nilpotent ideals; in view of Lemmas 7.3.2 and 7.3.3 R can have no nonzero nil ideals. By Theorem 6.1.1, $R[t]$ the polynomial ring in t over R is semisimple. Since we may assume that R satisfies a multilinear identity we have that $R[t]$ satisfies the identity of R. In other words we may assume that R itself is semisimple. By the corollary to Theorem 6.3.2 R satisfies a standard identity $[x_1, \cdots, x_r]$.

As a semisimple ring R is a subdirect sum of primitive rings R_α, α in some index set A, each of which is a homomorphic image of R and therefore satisfies $[x_1, \cdots, x_r]$. By Kaplansky's Theorem (Theorem 6.3.1) each R_α is a simple ring finite dimensional over its center.

Given $0 \neq a \in R$ let $T_a = \{\alpha \in A \mid a(\alpha) \neq 0\}$. Since R is a prime ring, given $a \neq 0$, $b \neq 0$ in R then $axb \neq 0$ for some $x \in R$. Hence $\varnothing \neq T_{axb} \subset T_a \cap T_b$. From this we see that we can enlarge the set $\{T_a \mid a \neq o \in R\}$ to a filter and hence to an ultra-filter $\mathfrak{F}$ on A. From our construction we have that R is isomorphically imbedded in $S = \pi R_\alpha / \mathfrak{F}$. By Theorem 7.3.1 S is a primitive ring and as we see by inspection it satisfies $[x_1, \cdots, x_r]$ since each R_α does. By Kaplansky's Theorem S is a simple algebra finite-dimensional over its center Z. Consider $RZ \subset S$.

An identity over the prime field P of Z satisfied by RZ is satisfied by R (if $P = GF(p)$ then $RP \subset R$, if P is the rational field clearing the denominators we get an identity with integers as coefficients which holds in R) hence in each R_α and so in $S = \pi R_\alpha / \mathfrak{F}$. By Theorem 6.3.3 we conclude that $RZ = S$.

Given $a \in R$ which is not a zero divisor in R it is not a zero divisor in RZ. If it were, pick $w \neq 0 \in RZ$ such that

$$w = b_1 \lambda_1 + \cdots + b_k \lambda_k, \; b_i \neq 0 \in R, \; \lambda_i \neq 0 \in Z$$

is of shortest "length" such that $aw = 0$. Now $a(wxab_1 - b_1xaw) = 0$ for any $x \in R$; but

$$wxab_1 - b_1xaw = (b_2xab_1 - b_1xab_2)\lambda_2 + \cdots$$
$$+ (b_kxab_1 - b_1xab)$$

is of shorter length than w and annihilates a hence $wxab_1 - b_1xaw = 0$ for all $x \in R$. Since $aw = 0$ we get $wRab_1 = (0)$ and so $w(RZ)ab_1 = (0)$. Since RZ is simple, being equal to S, and $w \neq 0$ we get $ab_1 = 0$. But a is not a zero divisor in R. This contradiction proves that a is not a zero divisor in RZ. Since RZ is a finite dimensional algebra the element a as a nonzero divisor in RZ must be invertible in RZ. We have shown that all regular elements in R are invertible in RZ.

By Lemma 7.3.2 R is a prime Goldie ring so has a ring of quotients $Q(R) = D_n$, the $n \times n$ matrices over a division ring D. We imbed $Q(R)$ in RZ as follows: given $x \in Q(R)$ then $x = a^{-1}b$ where a is regular in R and b is in R; map x into $a^{-1}b$ in RZ where a^{-1} is the inverse of a in RZ. Since $Q(R)$ is simple this is an isomorphic imbedding of $Q(R)$ in RZ. Since RZ is finite-dimensional over Z it satisfies a standard identity hence $Q(R)$ does. Thus D satisfies a standard identity; by Theorem 6.3.1 D must be finite-dimensional over its center. This completes the proof of Posner's Theorem.

Note that the theorem says that a prime ring satisfying a polynomial identity has a ring of quotients satisfying a polynomial identity. Actually more can be proved: if a semiprime ring satisfying a polynomial identity has a ring of quotients (it need not have such) then this ring of quotients satisfies a polynomial identity.

References

1. A. W. Goldie, The structure of prime rings under ascending chain conditions, *Proc. London Math. Soc.*, 8 (1958) 589–608.

2. ———, Semi-prime rings with maximum conditions, *Proc. London Math. Soc.*, 10 (1960) 201–220.

3. I. N. Herstein, Sul teorema di Goldie, *Rend. Accad. Lincei*, 35 (1963) 23–26.

4. E. Posner, Prime rings satisfying a polynomial identity, *Proc. Amer. Math. Soc.*, 11 (1960) 180–184.

5. Claudio Procesi, Sopra un teorema di Goldie riguardante la struttura degli anelli primi con condizioni di massimo, *Rend. Accad. Lincei*, 34 (1963) 372–377.

6. Claudio Procesi and Lance Small, On a theorem of Goldie, *J. of Algebra*, 2 (1965) 80–84.

CHAPTER 8

THE GOLOD-SHAFAREVITCH THEOREM

This last, and very short, chapter has one central result and some of its noteworthy implications. This result is due to Golod and Shafarevitch, published recently in a remarkable paper. Their theorem, rather easy to prove, provides a general method and technique for considering a large assortment of problems. As an immediate consequence of the main theorem one can construct a nil but not locally nilpotent algebra thereby giving a negative answer to the Kurosh Problem; one can construct a torsion group which is not locally finite thereby settling in the negative the general Burnside Problem. In addition the theorem gives rise to important results and examples in the theory of p-groups and in class field theory. However it seems likely that these successful uses of the method are merely a beginning, that a host of results await the application of the technique. We shall develop the results to such a point that we can construct the above-mentioned counterexamples to the Kurosh and Burnside Problems. We advise the reader to go to his original paper to see how the result is used in other algebraic areas.

Let F be any field and let $T = F[x_1, \cdots, x_d]$ be the polynomial ring over F in the d noncommuting variables $x_1, \cdots, x_d$. The algebra T has a degree function and is a graded algebra. We write T as $T = T_0 \oplus T_1 \oplus \cdots \oplus T_n \oplus$ where $T_0 = F$ and where T_n has as a basis the d^n elements $x_{i_1} x_{i_2} \cdots x_{i_n}$, where the x_{i_j} are chosen from $x_1, \cdots, x_d$. The elements of T_n are called the *homogeneous elements* of degree n.

Let $\mathfrak{A} = (f_1, f_2, \cdots)$ be the two-sided ideal of T generated by the homogeneous elements $f_1, f_2, \cdots$ of degree $2 \leq n_1 \leq n_2 \leq \cdots$ respectively. Furthermore let r_i be the number of n_2 which are equal to i.

Since $\mathfrak{A}$ is homogeneously generated the algebra $A = T/\mathfrak{A}$ inherits the grading of T; in fact $A = A_0 \oplus A_1 \oplus \cdots \oplus A_n \oplus \cdots$ where $A_i \approx T/(\mathfrak{A} \cap T_i)$. Let $b_n = \dim {}_F(A_n)$. The theorem about to be given furnishes a *sufficient* condition that A be *infinite dimensional* over F. This is the extremely important

THEOREM 8.1.1 (Golod-Shafarevitch). *For A as described above*

1. $b_n \geq db_{n-1} - \sum_{n_i \leq n} b_{n-n_i}$ *for* $n \geq 1$.
2. *if for each i the $r_i \leq [(d-1)/2]^2$ then A is infinite dimensional over F.*

Proof. We will exhibit linear mappings ϕ, ψ so that the sequence

$$(1). \ A_{n-n_1} \oplus \cdots \oplus A_{n-n_k} + \cdots \xrightarrow{\phi} A_{n-1} \oplus \cdots \oplus A_{n-1}$$

$$\xrightarrow{\psi} A_n \to 0$$

is exact, where the first sum runs over all $n_i \leq n$ and where the second sum is that of d copies of A_{n-1}.

Note that if we are able to do this then the inequality expressed in the theorem would be proved for then $db_{n-1} = b_n + \dim (\ker \psi)$ and since $\text{Ker } \psi$ is a homomorphic image of $\oplus_{n_i \leq n} A_{n-n_i}$ we would have $\dim (\text{Ker } \psi) \leq \sum_{n_i \leq n} b_{n-n_i}$, the net result of which would be $db_{n-1} \leq b_n + \sum_{n_i \leq n} b_{n-n_i}$, the desired conclusion.

Our objective then becomes that of defining the ϕ and ψ. To this end we shall first define mapping Φ and Ψ for the sequence

(2) $T_{n-n_1} \oplus \cdots \oplus T_{n-n_k} + \cdots$

$$\xrightarrow{\Phi} \overbrace{T_{n-1} \oplus \cdots \oplus T_{n-1}}^{d\text{-times}} \xrightarrow{\Psi} T_n \to 0$$

where Φ and Ψ are linear. Although the sequence will not be exact at the T-level it will turn out to be so at the A-level; that is, we shall induce the proper ϕ and ψ from these Φ and Ψ.

The mapping Ψ is defined simply by:

$$\Psi : t_1 \oplus \cdots \oplus t_d \to \sum_{i=1}^{d} t_i x_i \quad \text{for } t_i \in T_{n-1}.$$

To get Φ we proceed as follows: if

$$s_{n-n_1} \oplus \cdots \oplus s_{n-n_k} \oplus \cdots \in T_{n-n_1} \oplus \cdots$$
$$\oplus T_{n-n_k} \oplus \cdots$$

then $\sum s_{n-n_i} f_i \in T_n$. As an element in T_n we can write

$$\sum s_{n-n_i} f_i = \sum_{i=1}^{d} u_i x_i$$

where the u_i are *uniquely* determined elements in T_{n-1}. Define Φ by

$$\Psi : s_{n-n_1} \oplus \cdots \oplus s_{n-n_k} \oplus \cdots \to u_1 \oplus \cdots \oplus u_d$$

It is trivial that the Φ and Ψ defined are linear and have the proper ranges and domains. It is equally clear that Ψ is onto T_n so that the sequence (2) is exact at T_n.

Let $\mathfrak{A}_i = \mathfrak{A} \cap T_i$; our aim is to induce mappings ϕ and ψ from our Ψ and Ψ for the sequence (1).

If $t_1, \cdots, t_d \in \mathfrak{A}_{n-1}$ then since $\mathfrak{A}$ is an ideal of T, $\sum t_i x_i \in \mathfrak{A}$; by the properties of the grading $\sum t_i x_i \in T_n$. In short, it is in $\mathfrak{A}_n$. Thus the mapping Ψ induces down to $\psi : A_{n-1} \oplus \cdots \oplus A_{n-1} \to A_n \to 0$.

We now consider Φ. Suppose that $s_{n-n_1}, s_{n-n_2}, \cdots,$

s_{n-n_k}, $\cdots$ are in $\mathfrak{A}_{n-n_1}$, $\mathfrak{A}_{n-n_2}$, $\cdots$, $\mathfrak{A}_{n-n_k}$, $\cdots$ respectively. We must show that $u_1, \cdots, u_d$ defined by $\sum s_{n-n_i} f_i = \sum u_i x_i$ are in $\mathfrak{A}_{n-1}$. Since Φ is linear it suffices to do so for each s_{n-n_i} in $\mathfrak{A}_{n-n_i}$. Now

$$s_{n-n_i} f_i = \sum_{j=1}^{d} s_{n-n_i} g_{ij} x_j$$

where $f_i = \sum_{j=1}^{d} g_{ij} x_j$. Thus $u_j = s_{n-n_i} g_{ij}$ and so is in $\mathfrak{A}$ as s_{n-n_i} is in the ideal $\mathfrak{A}$. Being of the correct grade it is in $\mathfrak{A}_{n-1}$. Therefore Φ too induces down to a map ψ from $A_{n-n_i} \oplus \cdots \oplus A_{n-n_k} \oplus \cdots$ to $A_{n-1} \oplus \cdots \oplus A_{n-1}$.

We must still show exactness at $A_{n-1} \oplus \cdots \oplus A_{n-1}$. We first establish that $\phi\psi = 0$. If $s_{n-n_1}, \cdots, s_{n-n_k}$ are in $T_{n-n_1}, \cdots, T_{n-n_k}$ respectively then

$$(s_{n-n_1} \oplus \cdots \oplus s_{n-n_k})\Phi\Psi = \sum_{i=1}^{d} u_i x_i$$

where $\sum u_i x_i = \sum s_{n-n_i} f_i$; since the f_i are in $\mathfrak{A}$ the sum, $\sum s_{n-n_i} f_i$, is in $\mathfrak{A}$ hence $\sum u_i x_i \in \mathfrak{A}$. In other words, $\Phi\Psi$ maps $T_{n-n_1} \oplus \cdots \oplus T_{n-n_k} \oplus \cdots$ into $\mathfrak{A}$ hence $A_{n-n_1} \oplus \cdots \oplus A_{n-n_k} \oplus \cdots$ is mapped into 0 by $\phi\psi$.

We must still show that if $(t_1 \oplus \cdots \oplus t_d)\Psi \in \mathfrak{A}$ then we can find elements $u_1, \cdots, u_d$ in T_{n-1} such that $t_i - u_i \in \mathfrak{A}$ for $i = 1, 2, \cdots, d$ and such that $\sum u_i x_i = \sum s_{n-n_i} f_i$ for some s_{n-n_i} in the appropriate T_{n-n_i}. Suppose then that $(t_1 \oplus \cdots \oplus t_d)\Psi = \sum t_i x_i \in \mathfrak{A}$; being in $\mathfrak{A}$, which is generated as a two-sided ideal by the f_j we have that

$$\sum t_i x_i = \sum a_{kq} f_q b_{kq} + \sum c_q f_q$$

where the a_{kq}, b_{kq}, c_q are homogeneous and where the degree of b_{kq} is at least 1. On comparing degree on both sides we may even assume that the $a_{kq} f_q b_{kq}$, $c_q f_q$ are all in T_n. Since the b_{kq} are of degree at least 1, b_{kq}

$= \sum_{m=1}^{d} d_{kqm} x_m$ hence

$$\sum a_{kq} f_q b_{kq} = \sum a_{kq} f_q d_{kqm} x_m = \sum d_m x_n$$

where $d_m = \sum_{k,q} a_{kq} f_q d_{kqm}$. But since $f_q \in \mathfrak{A}$ we have that $d_m \in \mathfrak{A}$. If we write $\sum c_q f_q = \sum_{j=1}^{d} u_i x_i$ we then have that

$$\sum_{i=1}^{d} t_i x_i = \sum_{i=1}^{d} d_i x_i + \sum_{i=1}^{d} u_i x_i$$

hence $t_i - u_i = d_i \in \mathfrak{A}$. But $(c_1 \oplus \cdots \oplus c_i \oplus \cdots) \Phi = u_1 \oplus \cdots \oplus u_d$ by the definition of Φ; hence we have proved the exactness of (1) at $A_{n-1} \oplus \cdots \oplus A_{n-1}$. This proves part (1) of the theorem.

We now consider part (2). For formal power series in t with integer coefficients we declare

$$\sum_{n=0}^{\infty} c_n t^n \geqq \sum_{n=0}^{\infty} d_n t^n$$

if $c_i \geqq d_i$ for all i.

From part (1) we have that

$$\sum_{n=1}^{\infty} b_n t^n \geqq \sum_{n=1}^{\infty} d b_{n-1} t^n - \sum_{n=1}^{\infty} \sum_{n_i \leqq n} b_{n-n_i} t^n.$$

From the definition of r_i we can write

$$\sum_{n_i,m} b_m t^{n_i+m} = \sum t^{n_i} \sum_m b_m t^m = \left(\sum_{i=2} r_i t^i \right) \left(\sum_{m=0} b_m t^m \right).$$

Let $P_A(t) = \sum_{m=0}^{\infty} b_m t^m$. The above relations become:

$$P_A(t) - 1 \geqq dt P_A(t) - \left(\sum_{i=2}^{\infty} r_i t^i \right) P_A(t);$$

therefore

$$P_A(t) \left(1 - dt + \sum_{i=2}^{\infty} r_i t^i \right) \geqq 1.$$

Now if the coefficients in the formal power series expansion of

$$\left(1 - dt + \sum_{i=2}^{\infty} r_i t^i\right)^{-1}$$

are nonnegative we get that

$$P_A(t) \geqq \left(1 - dt + \sum_{i=2}^{\infty} r_i t^i\right)^{-1}$$

and so, an infinite number of the b_n must be different from 0. Hence A is infinite dimensional.

This itself is of great interest and we single it out as a theorem before completing the proof of part (2) of Theorem 8.1.1.

THEOREM 8.1.2. *A is infinite dimensional over F if the coefficients in the power series expansion of*

$$\left(1 - dt + \sum_{i=2}^{\infty} r_i t^i\right)^{-1}$$

are nonnegative.

To finish the proof of Theorem 8.1.1 we leave it to the reader to show that if each $r_i \leqq [(d-1)/2]^2$ then the criterion of Theorem 8.1.2 is satisfied and so the theorem is established.

From Theorem 8.1.1 we can readily construct a finitely generated nil algebra which is not nilpotent, hence settling the Kurosh Problem in the negative. The example we give is not the best possible; our restriction on the countability of F can be removed and we can get by with two generators instead of three. But one example suffices. Hence

THEOREM 8.1.3. *If F is any countable field there exists an infinite dimensional nil algebra over F generated by three elements.*

Proof. Let $T = F[x_1, x_2, x_3]$ be the free algebra over F in the three noncommuting variables x_1, x_2, x_3. Since T is graded we can write $T = F \oplus T_1 \oplus \cdots \oplus T_n \oplus \cdots$ where T_i is homogeneous of degree i. The ideal $T' = T_1 \oplus T_2 \oplus \cdots \oplus T_n \oplus \cdots$ of T is countable and we enumerate its elements as s_1, s_2, $\cdots$. Pick $m_1 \geqq 2$ and let $s_1{}^{m_1} = s_{12} + s_{13} + \cdots + s_{1k_1}$ where $s_{1j} \in T_j$. Now pick an integer $m_2 > 0$ so that $s_2{}^{m_2} \in T_{k_1+1} \oplus T_{k_1+2} \oplus \cdots$, hence $s_2{}^{m_2} = s_{2,k_1+1} + \cdots + s_{2,k_2}$ where $s_{2j} \in T_j$. Continue in this pattern. Hence we have chosen integers $m_i > 0$ and $k_i < k_2 < \cdots$ such that $s_i{}^{m_i} = s_{i,k_{i-1}+1} + \cdots + s_{i,k_i}$ with $s_{ij} \in T_j$. Let $\mathfrak{A}$ be the ideal of T generated by all the s_{ij}. By our choice of the s_{ij} the integers r_k in Theorem 8.1.1 are all at most 1. Since $d = 3$, $r_i \leqq 1 \leqq [(d-1)/2]^2$ holds true, therefore applying Theorem 8.1.1 we get that $T/\mathfrak{A}$ is infinite dimensional. Since $\mathfrak{A} \subset T'$ we have that $T'/\mathfrak{A}$ is infinite dimensional over F. By construction $T'/\mathfrak{A}$ is a nil algebra. Since it is generated by three elements $T'/\mathfrak{A}$ is the required example.

We close the chapter (and the book) by settling the general Burnside Problem in the negative.

THEOREM 8.1.4. *If p is any prime number there exists an infinite group G generated by three elements in which every element has finite order a power of p.*

Proof. Let F be the prime field with p elements and let $\mathfrak{A}$ be the ideal in $T = F[x_1, x_2, x_3]$ constructed in the course of proving Theorem 8.1.3. Let $A = T/\mathfrak{A}$ and let a_1, a_2, a_3 be the elements $x_1 + \mathfrak{A}$, $x_2 + \mathfrak{A}$, $x_3 + \mathfrak{A}$ respectively. Let G be the multiplicative semigroup in A generated by the elements $1 + a_1$, $1 + a_2$, $1 + a_3$. Any element in G is of the form $1 + a$ where $a \in T'/\mathfrak{A}$ (so is a nilpotent). For large enough n, $a^{p^n} = 0$ hence $(1+a)^{p^n} = 1 + a^{p^n} = 1$ since we are in characteristic p. Hence G is a group—in fact a torsion group—and every element of G has order a power of p. We claim that G is infinite. For if G is finite the linear combinations of its elements would form a finite dimensional algebra B over F; since $1, 1+a_i$ are in G the element $a_i = (1+a_i) - 1 \in B$. Since 1, a_1, a_2, a_3 generate A we get $A = B$ contradicting that A is infinite dimensional over F. This finishes the proof.

References

1. E. S. Golod, On nil algebras and finitely approximable groups, (Russian) *Izv. Akad. Nauk SSSR.*, Ser. Mat. 28 (1964) 273–276.

2. E. S. Golod and I. R. Shafarevitch, On towers of class fields, *Izv. Akad. Nauk SSSR.*, Ser. Mat. 28 (1964) 261–272.

SUBJECT INDEX

NAME INDEX

198